HUMAN DIVERSITY

HUMAN DIVERSITY

�֍

Alexander Alland, Jr.

ANCHOR BOOKS

ANCHOR PRESS/DOUBLEDAY

Garden City, New York

1973

Alexander Alland is Associate
Professor of Anthropology
at Columbia University

Human Diversity was originally published by Columbia University in 1971. The Anchor edition is published by arrangement with Columbia University Press.

Anchor Books edition: 1973

Copyright © 1973 by Alexander Alland, Jr.
Copyright © 1971 by Columbia University Press
ISBN: 0-385-08022-0
Library of Congress Catalog Card Number 72–96371
Printed in the United States of America

TO MY WIFE

Sonia

AND MY CHILDREN

David & Julie

PREFACE

SCIENCE is value free, but scientists are not. Individual research choices are determined by personal factors which sometimes include political beliefs. Such an influence is natural and can orient research toward socially significant results. It is harmful only when ideology limits or destroys the objectivity of the research process.

Race is one of the most emotion-laden topics in social and biological science. It is impossible to approach this subject without preconceived notions and political bias. Much of the published material on race is either ambiguous or laden with half-truths which can only confuse lay readers. The concept of race as it has been used by the majority of writers in and out of anthropology is largely a folk category, more often than not accepted uncritically. Yet, as long as race occupies a place in economic and political life, it must be discussed. To ignore it would be to ignore empirical data on how our society works.

I have chosen to write on this subject for two reasons, the first of which is my belief that scientists must concern themselves with social issues that lie within their professional competence. The second reason has more to do with my own place within anthropology. I have always been interested in questions concerning real and supposed links between biology and culture. I have taught courses in physical and medical anthropology as well as ethnology and have tried in each to search out possible relationships betwen biology and human behavior. The whole question of the evolution of man as a distinct species revolves around this problem. I have also been critical of those who, in my opinion, have been hasty in accepting race as more than an item of folk taxonomy, particularly those who link race to behavior as manifested in a society's tradition. The publications of such authors as Robert Ardrey and Arthur Jensen force me to speak out against distortions concerning both the biological basis of human behavior and what it is professional anthropologists do and say. Let it be clear that few if any of us deny a role to biology in the shaping of individual differences, nor are we egalitarians in the sense that we believe that all men are born equal. Our national values have taught us that all men should be given equal opportunity *not* that they are born equal. Our science has taught us that the methods and results of studies demonstrating individual differences cannot be extrapolated to group differences, that such categories as race may be social rather than biological entities, and that concepts like intelligence should not be reified.

I wish to thank my students and my wife, Sonia, for inspiration; Marvin Harris and Ralph Holloway for helpful comments on the manuscript. Acknowledgements are gratefully made to a number of publishers and individuals for permission to quote extensively from their works. They include *American Anthropologist; Eugenics Quarterly;* The Free Press; *Harvard Educational Review;* Holt, Rinehart and Winston; McGraw-Hill; *Scientific American;* R. Sitgreaves; and

Wayne State University Press. The bibliographical details for all quotations are given in the text and in the Bibliography. The figures on pages 8, 9, and 12 are from *Evolution and Human Behavior,* Copyright © 1967 by Alexander Alland, Jr.; published by the Natural History Press of Doubleday & Co., Inc.

Alexander Alland, Jr.

CONTENTS

HUMAN DIVERSITY

Chapter 1

❈

INTRODUCTION

HUMAN BEINGS (*Homo sapiens*) are the most widely distributed of animal species. Archeological and fossil evidence suggests, however, that our pre-sapient ancestors lived only in tropical and subtropical areas. The full exploitation of environmental space occurred with the evolution of culture, an adaptation rooted in man's genetic background but which operates beyond genetic control.

Penetration into new environmental zones exposed human populations to differential selective forces which operated on both genetic and cultural systems. Pressure on the genetic system was probably severest when culture was most rudimentary, for advances in technology tend to buffer man from the strictly natural aspects of his environment. By the time of major expansions into colder climates this buffer had developed to the point where, given the time limits involved, genetic isolation would no longer produce speciation. Divergences from the basic human pattern were minimized. Wherever man went, his most characteristic adaptation, the capacity for culture, remained optimal. Major adaptive changes occurred in culture, a nonsomatic, nongenetic system, rather than in genetic structure.

The genetic integrity of the human species as a whole has been preserved also through migration and inbreed-

ing, common features of human existence. There is
some evidence, too, that certain aspects of physiological
and chemical evolution in primates have tended to slow
down those biological processes which promote genetic
diversity.

This is not to say that all men are biologically equal
or that all populations contain exactly the same genetic
material in equal frequencies. It is clear that those biological forces which drive populations onto different
genetic paths continue to operate. The existence of
culture, however, distorts the process of differentiation
to the extent that comparison with nonhuman populations is often more misleading than revealing. Culture
does not exclude man from the biological realm, far
from it, but it does exert its special influence on the
entire adaptive process. Thus, while the mechanisms
which lead to genetic change are well known for all
species, including man, anthropologists are generally
unclear as to which ones are responsible for specific
biological configurations.

Until recently the problem has been compounded because attention was directed toward the wrong type of
solution. Quite naturally anthropologists began their
analytic task by cataloging data on the basis of what
were taken to be natural units: cultures for the ethnolo-
gist, and populations for the physical anthropologist.
Next, in order to simplify and generalize the data, attempts were made to group these units into higher-
order and relatively noncontinuous categories: culture
areas on the one hand, and races on the other. This
task turned out to be impossible. No clear set of criteria
for classification was ever agreed upon and the boundaries offered by different authorities diverged widely.
The failures occurred because in most cases biological

and cultural variation is continuous; natural groupings rarely occur.

Population genetics proved long ago that units below the level of the species could not be taken as discrete typological entities, separate in kind from other such entities. Races represented different clusters of gene frequencies and were defined as overlapping units which differed primarily in mean gene frequencies. So-called typological thinking gave way to populational thinking, but race was still seen as a useful if vague unit lying somewhere between the species and the population. More recently, biologists have questioned the utility of race altogether. Evidence from infrahuman species as well as from human populations tends to show that no successful units can be created on the subspecific level. The data have forced research on human diversity onto a new path. Classification for the sake of classification has given way to the analysis of process.

Early attempts at cataloging were not completely without value, however. An analysis of the distributions of cultural and biological groupings shows little correspondence between them. The evidence for this is overwhelming and definitive. It proves that major behavioral differences among human groups are due not to biological variables but to culture. Yet for many the racial myth persists. It provides a useful if inaccurate explanation for centuries of political, social, and economic exploitation.

This book is an introduction to the study of human diversity in its biological and sociological aspects. Its major argument is that race as a concept is valid only in sociological discourse. This idea is not original with me. A new generation of physical anthropologists and biologists has come to question the utility of race as a taxonomic device.

Those who retain the term in human biology have usually retreated to the point at which races and populations are equated. If this is the case, why retain the term "race"? Its folk connotation is so far removed from the populational concept, which will be discussed below, that its continued use by scientists only confuses the issue.

In sociology and in social life, on the other hand, race is a real phenomenon, although its meaning varies according to who is using the term, when it is being used, and what group it is being applied to. As a concept which defines groups partly on the basis of real or imagined biological characteristics, race retains an almost unique force in the shaping of human affairs.

An analysis of the biological side of this problem must deal with the extreme view that all human behavior is controlled exclusively by culture. Recent advances in behavioral genetics have demonstrated that many individual behavioral differences are due in part to specific genetic mechanisms. There is also some inferential evidence that populational differences might occur in the distribution of genes which influence behavior. What needs stressing is the fact that the background provided by genes affects only the probability that certain types of behavior will occur. These probabilities, as such, cannot be determinants of behavior. That is to say, the effect of environment, including culture, may override and suppress or, conversely, accentuate certain features of genetic structure.

The plan of this book is simple. It begins with a short introduction to genetics for the reader with no biological training. Forces which produce human diversity are then discussed in Chapter 3, and reasons for abandoning the concept of race in biology are presented. Chapter 4 is concerned with the notion of Caucasoid antiquity and theories which link fossil man to specific living groups. The recent work of Carleton Coon, the best-known spokesman for those who would retain the concept of race in biology, is examined. Coon argues for the antiquity and stability of what he considers to be the living races of man. No examination of the race question could be complete without a discussion of his contribution. The reader is to be forewarned that I am highly critical of Coon. For the sake of fairness, I suggest that his works be consulted in detail.

Chapter 5 examines behavioral genetics in relation to problems of human diversity. It does not present a complete review of experimental or theoretical work in this field, but is instead set in the specific perspective of racial studies. The question of intelligence differences, which figures in behavioral genetics, is reserved for the last chapter.

Chapter 6 deals with race as a sociological phenomenon. The idea that racial concepts are mainly projections of folk ideology is criticized. It is argued that racial myths develop in specific historical contexts involving encounters between social groups. Such myths operate to maintain social distance between groups for socioeconomic reasons.

The last chapter is concerned with the question of racial intelligence and includes a critique of the recent work by Jensen. In my opinion the great debate over inherent group differences serves best as an illustration of the confusions which arise when a biological label is applied to a sociological phenomenon. Chapter 7 is, therefore, meant as a synthesis drawing the major arguments of the book together.

I write as an anthropologist trained in ethnology, but one who teaches and maintains a strong interest in physical anthropology, that branch of my science which deals with human biology. I hope that I bring a double awareness to the questions raised in this book, for they span two fields. I not only have attempted to compare the sociological and biological views but also have taken a stand in defense of those who would include behavioral genetics as a vital field in the search for explanations of man's evolution as well as his contemporary behavior. Behavioral genetics is a new and exciting field. It has suffered greatly from attacks by extreme culturologists and misapplications by those who would oversimplify the relationship between environment and genes. It has also been seized upon by those who would like to prove that racial differences are real and more than skin deep. The work of competent behavioral geneticists stands as a corrective to both schools. It contributes to our understanding of the full complexity of human behavior set in the context of human diversity.

Chapter 2

⚘

BASIC GENETICS

THE PURPOSE of this chapter is to acquaint readers who have no background in biology with some of the basic material of genetics. Those who are familiar with such material may wish to skip directly to the next chapter.

Genetics and Chemistry. Inheritance in all living things (except some simple viruses) is controlled by a complex molecule known as deoxyribonucleic acid, or DNA. It consists of a pair of intertwined chains of indefinite length composed of sugar phosphate molecules. The two chains are linked together at regular intervals by hydrogen bonds which come off of one of four bases. These bases—adenine, thymine, cytosine, and guanine—can be arranged in any order along one of the two chains. But guanine can link only with cytosine and adenine can link only with thymine. Thus, the sequence on one chain determines the sequence on the other. These four bases serve as units of information, the letters of an alphabet, which direct the synthesis of proteins vital to the cell's metabolism and development. Each protein consists of a specific combination of amino acids arranged in a particular order. Three letters in the code, a series of three bases, spell a single amino acid. A definite series of such three-letter words, arranged in specific order on a part of

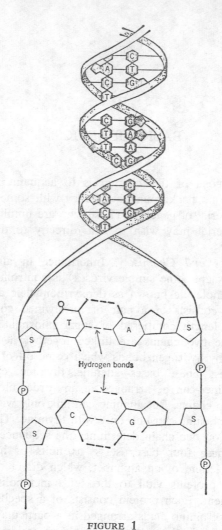

Hydrogen bonds

FIGURE 1
*A schematic diagram of the Watson-Crick model
of the DNA molecule.* (*Alland 1967*)

the DNA chain, constitutes a sentence which translates either a complete protein or a complete chain of a multiple chain protein. In a process which is too complicated to describe here, specific segments of the DNA molecule control the production of proteins within the cell.

DNA is stored in the nucleus of the cell, in structures

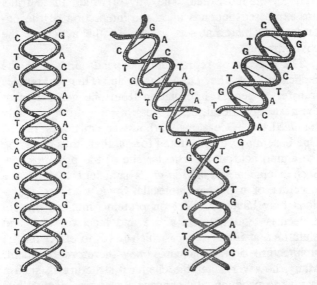

FIGURE 2

The double-helix model of the gene and chromosome structure (left) and of its replication (right) as suggested by the Watson-Crick hypothesis. (Alland 1967)

known as chromosomes. The number of such chromosomes in each organism of a particular species is constant, but the number of chromosomes among species is highly variable. Human beings, for example, have forty-six chromosomes representing a set of twenty-three homologous pairs. This *diploid* number represents the genetic contribution of an individual's parents, each of whom contributes a haploid set of twenty-three chromosomes. Each parent (the male in the sperm, the female in the egg) contributes one-half the chromosome complement responsible for the future hereditary traits of an individual. The way in which these traits are expressed depends upon the interaction of the paternal and maternal sets as well as the action of the environment.

The process of reproduction depends first upon the replication of the DNA molecule. This guarantees transfer of genetic information from one generation to the next. In general, the process of replication is highly standardized. Occasionally, however, mistakes occur. One base may be substituted for another, thus changing the amino acid at a particular site of the protein code. Such a change, or *mutation,* is sufficient to alter the structure of a protein molecule. Some of these mutations may have no effect on protein function; protein molecules tend to have active and inactive sites, and mutations at inactive sites which have no effect on the architecture of the molecule may therefore be silent. Many, however, alter the chemical structure in such a way as to produce vital changes in the organism. While most mutations tend to be harmful because they represent a change from an evolved integrated system, some are beneficial; that is, they increase the efficiency of the organism.

Genes. Those segments of the DNA molecule which are responsible for the production of a complete protein molecule or a complete segment of a complex protein molecule are referred to as genes. The morphological and physiological traits which one observes in an individual depend ultimately upon these genes, but they do not produce traits directly. An individual organism is a product of genetic structure, known as the *genotype,* and environmental interaction with this structure. The outcome of this process of interaction between heredity and environment, the *real* organism, is known as the *phenotype.*

Phenotypes may be highly variable even in species with little genetic variation, or they may be rather uniform even in those species which encompass a good deal of genetic or genotypic variation. This is possible because genetic mechanisms exist which are capable of masking or preserving genotypic variation and because the environment may in one way or another suppress existing genotypic variation.

Genetic Change. In asexually reproducing organisms, like many bacteria, most genetic change depends upon mutation. In sexually reproducing species, on the other hand, genetic material may be recombined from generation to generation. Since each parent contributes one-half of his or her genetic information (the haploid number of chromosomes), and since this half is made up at random from that individual's own parental generation (through independent assortment), a good deal of genetic recombination usually takes place. Such recombination can produce new genotypes and phenotypes without any changes in individual genes and without any change in chromosome structure. In addition, during the process of meiosis (the formation of sexual cells, eggs or sperms) segments of homologue chromosomes may cross over one another, become detached from their original chromosome, and reattach themselves to the opposite homologue. Finally, chromosome

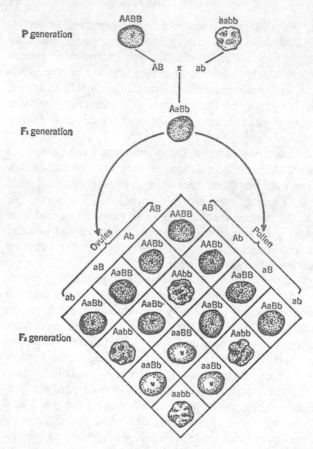

FIGURE 3

An example of Mendel's law of independent assortment, in which strains of peas with yellow and smooth seeds and with green and wrinkled seeds are crossed. A and a represent yellow and green, and B and b smooth and wrinkled surfaces, respectively.

(*Alland 1967*)

structure may change due to various types of breakage and relinkage to nonhomologues. Such processes increase the possibility of genetic variation.

The Locus. The segment of a particular chromosome responsible for a particular trait is known as a locus. There may be, for example, an eye color locus, or a hair form locus, or a locus responsible for the production of a particular blood group such as A, AB, B, or O. Since mutations can occur at any locus, variant forms for the same trait may be present in the species. Variant genetic units which are members of a particular trait class and which occur at the same genetic locus are called alleles. In the ABO(H) blood system of humans three such alleles exist at what is known as the I locus. These are A, B, and O. An individual will receive one of these three alleles from each parent. Combinations of these alleles result in one of the following possible genotypes: AA, AO, AB, BB, BO, and OO. These six combinations produce only four phenotypes: A, AB, B, and O. These phenotypes occur because certain alleles may mask others and because certain alleles may act in combination to produce a third trait. When one allele is capable of masking the effects of another, it is said to be *dominant;* the masked allele is *recessive.* When alleles act in combination to produce a third (sometimes but not always intermediate) trait, they are referred to as *codominant.* These are relative terms. An allele is dominant or recessive in relation to some allele. In the ABO system A and B are dominant over O but are codominant to each other. The only way for blood type O to appear is for an individual to receive an O allele from each of his parents. An individual carrying the same alleles at a single locus is referred to as *homozygous* at that locus. An individual carrying different alleles at the same locus is said to be *heterozygous* at that locus. Individuals with the same phenotype may have different genotypes. People who are blood group A, for example, may be either AA homozygous or AO heterozygous.

The ABO system in human beings depends upon a single locus. Many traits, however, are produced at various loci which may be located either on a single chromosome or even on different chromosomes. The hemoglobin molecule depends upon at least two different loci, and skin color in humans is dependent upon at least three separate loci.

While genetic units can never mix, they may act in combination to produce a wide range of intermediate effects. Thus, in mixed populations one may find a gradient of skin color from white through various shades of brown to very dark brown so often referred to as black. One must remember also that phenotypes are the result of genetic-environment interaction. Such interactions often increase the range of variation in a population. Stature in human beings, for example, is probably the result of polygenetic and environmental effects.

Genes and Environment. One of the major problems in genetics is the separation of environmental and genetic factors. Sibling and offspring-parent crosses, plus controlled environmental conditions, make genetic analysis possible. Human genetics presents special problems since people take a long time to breed, often do not keep breeding records, are not brought up under controlled environmental conditions, and cannot, for obvious reasons, be crossed sibling to sibling or backcrossed offspring to parent. Geneticists interested in human beings rely on what geneological information is available, statistical studies of trait distributions in populations, and identical twin studies. Identical twins reared apart are particularly valuable because they present an analogue to experimental conditions in which hereditary information is held constant while the environment is varied. Differences between such individuals are probably due to environmental factors; similarities, to genetic factors.

While studies of continuous variation in human populations are common, it is difficult to sort out genetic from environmental causes or to determine how many genes are operating in the system. Single gene traits are considerably easier to study. They can often be inferred from the examination of genealogies. Human geneticists and physical anthropologists have turned more and more to the study of various blood group systems, other elements found in blood serum, and such traits as the ability to taste PTC, fingerprint patterns, and types of ear wax. These are frequently under the control of relatively simple genetic units and can be analyzed on the chemical level in the laboratory.

Selection. Natural selection (the process by which evolution operates) depends upon the unequal capacities of organisms to reproduce. The relative success of a particular trait variant is measured in terms of a selective coefficient: the relative reproductive disadvantage of organisms bearing a trait in comparison with those not bearing the trait.

Genes with selective coefficients less than one (an arbitrary number assigned to the gene associated with higher reproductive success) are reduced in frequency through natural selection. Elimination and retention of genes, however, are statistical processes. That is, we can never be exactly sure whether or not a particular organism will produce its full potential of offspring or, conversely, if a disadvantaged organism will actually be eliminated from the population before it reproduces.

The frequency of recessive genes is always considerably higher than the homozygous recessive phenotype. It is therefore difficult, if not impossible, to remove an "undesirable" recessive trait from a natural population. Selective coefficients predict the expected frequencies of genes in comparison to one another from generation to generation. The statistical expectation that a group of organisms of a given genotype will produce a higher or lower number of offspring in comparison to the average for the population is referred to as fitness. In cases where there is no environmental or genetic change (a hypothetical situation) gene frequencies will come to equilibrium on the basis of the Hardy-Weinberg law, based on the formula $(p + q)^2 = 1$, or $p^2 + 2pq + q^2 = 1$ where p refers to the frequency of allele A (dominant) and q to allele a (recessive). In the case of multiple alleles the formula can be expanded. For the ABO(H) system, for example, the Hardy-Weinberg formula would be $(p + q + r)^2$ where p referred to the frequency of A, q to the frequency of B and r to the frequency of O.

Mutation is a continuous process, but mutation rates are usually constant so that they can be worked into the equilibrium model. When selection pressures change, gene frequencies will change at a predictable rate and, if new pressures are constant over a span of generations, a new genetic equilibrium will be established.

The adaptation of a species to the environment depends in part upon the maintenance of efficient genetic structures from generation to generation. Improvements in the adaptation of a species depend upon changes in the phenotypic structure often through modification of the genetic system. Normal replication of the DNA molecule preserves continuity while accidental mutations and other genetic changes provide the variation necessary for adaptation.

Two separate streams of events contribute to the evolutionary process. One of these is genetic change. The other includes selective environmental factors which act independently of genetic change. Mutation does not arise in response to environmental demand nor does the environment in some way become modified to accommodate a species. There is, therefore, no teleology in evolutionary theory.

With some glaring exceptions species are natural units which reproduce within themselves but which normally, i.e., under natural conditions, do not cross-fertilize with other such units. Under most conditions selection operates on local breeding populations which are the "evolutionary units" of the species. This is because most species are widely distributed in space and are segmented into populations which exchange genetic material in much higher frequency within the unit than among units. The gene pools of local populations vary in relation to local genetic, environmental, and accidental events. Natural selection acts on these populations in slightly different ways, but successful variations which develop in one population may be passed on to other populations of the same species through interbreeding in a process known as gene flow.

Because of its wide geographical distribution the human species is divided into a tremendously large number of breeding populations. Each has a slightly different environment and each a slightly different genetic history. In general, however, genetic differences be-

tween such populations consist of different frequencies
of the same alleles. Only rarely is a particular allele the
exclusive property of one or another population. In
terms of genetic distribution, populations may be con-
ceived of as overlapping statistical entities rather than
as discrete, biologically closed units.

Gene frequencies depend upon mutation rates
(which are always very low), the effect of natural selection, and two nonselective factors: interbreeding and
genetic drift (random change). Under *theoretical* conditions, if the environment is stable, the population infinite, with neither mutation nor selection operating,
gene frequencies will remain constant from generation
to generation. When mutation rates are constant and
when natural selection exerts a constant pressure, these
two interacting forces will bring gene frequencies into
equilibrium. If, however, different populations interbreed, their separate gene pools will combine to produce a new set of gene frequencies which will then
come to equilibrium. Natural selection may operate on
this changed gene pool, which, of course, then disturbs
the equilibrium anew.

Nonselective or accidental change can have a profound influence on gene frequencies. If a segment of a
population migrates away from its home territory, it is
likely to differ genetically from its parent population by
virtue of nonrandom sampling. The smaller the original
population and/or the smaller the migrant population
the greater the chance that the two groups will diverge.
Such divergences may also occur in situ from generation to generation. Accidental changes in such populations may significantly shift gene frequencies. Hurricanes, volcanoes, severe storms on islands, famine and
flood, epidemics against which there is no genetic im-

munity, may all bring about such shifts in the gene pool. Any change in gene frequency which is due to accidental events is called *genetic drift*. In addition, certain social factors may affect fertility. Celibacy, social class, and marriage rules may all tend to skew the gene pool in directions which are out of balance with biological processes. Assortative mating (mating in which individuals choose their sexual partners on some nonrandom basis) can also produce genotypic variation. It is possible that assortative mating among humans has affected genetic structure, but such an idea must be treated as a hypothesis.

In the ongoing process of evolution, drift and natural selection can operate in sequence. That is, when the gene pool upon which natural selection operates is changed by random factors, selection may take a new direction. Evolution is an opportunistic process; it works on what is there, and if what is there is changed, the direction of natural selection may also change.

Human populations rarely, if ever, remain rooted to the same spot. Over a long time span migrations occur (some over relatively short distances, others quite long), particularly among groups on the hunter-gatherer level of technological development. This occurs particularly when climatic change alters the environment sufficiently to affect the game supply. Population movement, as already pointed out, can produce random changes in genetic structure, but migration also implies a change in environment. Thus, when a population moves from one geographic area to another, natural selection pressures may also change, bringing about a shift in gene frequencies.

Since many organisms are territorial (particularly birds, fish, many ungulates, some carnivores, and some primates) or for environmental reasons are limited in their contact with other populations, rates of interbreeding may be fairly low. In man, interbreeding has probably constituted a part of history from early times. When differences are found in gene frequencies among populations of the same species, these differences may be due to mutation, to selection, to drift, to interbreeding, to assortative mating, or combinations of these. In addition, shifts in the gene pool may be produced as a result of the migration to some new environmental setting.

Note that I have referred to changes in gene frequencies in populations. Populations may diverge from one another at one point in time, but they may merge again later. Populations distinctive in some genetic traits may disappear totally from the species without appreciably changing that species as an overall unit. Populations in the temporal sense are ephemeral units rather than fixed entities.

Chapter 3

❀

RACE AND BIOLOGY

ONE OF the curious facts about the term "race" is the ambiguity which adheres to it. The *Oxford English Dictionary* offers the following set of definitions (obsolete definitions omitted):

I. A group of persons, animals, or plants, connected by common descent or origin
    1. The offspring or posterity of a person; a set of children or descendants
        b. Breeding, the production of offspring
    2. A limited group of persons descended from a common ancestor; a house, family, kindred
        b. A tribe, nation, or people regarded as of common stock
        c. A group of several tribes or peoples, forming a distinct ethnical stock
        d. One of the great divisions of mankind, having certain physical peculiarities in common
    3. A breed or stock of animals; a particular variety of a species
        b. A stud or herd (of horses)
        c. A genus, species, kind of animals
    4. A genus, species, or variety of plants
    5. One of the great divisions of living creatures:
        b. Mankind
        c. A class or kind of beings other than men or animals

    d. One of the chief classes of animals (as beasts, birds, fishes, insects, etc.)

6. Without article:
    b. Denoting the stock, family, class, etc. to which a person, animal or plant belongs . . .
    c. The fact or condition of belonging to a particular people or ethnical stock; the qualities, etc. resulting from this

7. Natural or inherited disposition

II. A group or class of persons, animals, or things, having some common feature or features

8. A set or class of persons
    b. One of the sexes

9. A set, class, or kind of animals, plants or things

10. A particular class of wine, or the characteristic flavor of this . . .
    b. Of speech, writing, etc.: A peculiar and characteristic style or manner, *esp.* liveliness, sprightliness, piquancy

Scientists concerned with human variation have not done much better in constructing a workable definition of race. Ashley Montagu, for example, in his *Introduction to Physical Anthropology* (3rd edition, 1960) drops the term altogether, substituting "major group" for "race" and "ethnic group" for "subrace." This is done in an attempt to dissociate a scientific concept from emotional contamination. Unfortunately, however, the term "ethnic group" itself has sociological connotations which are unrelated to biological entities.

Montagu's so-called major groups are also rather ambiguous, although the ambiguity, in this case, *is* intentional. Montagu does not accept rigid biological subgrouping for the human species. The Negroid major group is described as follows: "Among the Negroids the skin is typically dark brown but is often black, and even yellowish-brown in some groups. The head hair varies from partly curled to pepper corn in form. . . . The head is long, the nose is broad and flat with wide nostrils, the ears small, and there is some prognathism (forward projection of the upper jaw), and the lips are thick and everted." (1960:419–20)

Under the Caucasoid major group we find:

This division of mankind is often called "white." The term is not an accurate one for the reason that it includes many people of dark skin color. The reason for giving this major group the name "Caucasoid" originates in the choice made by Blumenbach, the father of Physical Anthropology, who in the late eighteenth century described and named the type from a female skull, whose beauty had much impressed him, and came from Georgia in the Caucasus and which seemed typical of the cranial characters of the group. . . . Head hair is usually wavy, but ranges from silky straight to various degrees of curliness. It is almost never woolly, rarely frizzly, and is seldom as coarse or sparsely distributed as in mongoloids. . . . Skin color varies from white to dark brown. All forms of head shape occur . . . (448)

Montagu defines thirteen different "ethnic" groups under the Caucasoid major group. However, his defining characteristics intentionally make it impossible to sort individuals neatly into categories. Under Mediterranean, for example, he states: "Skin color varies all the way from tawny white to light or medium darkbrown, . . . hair is dark brown or black, and varies from a light wavy form to a loose curl; . . . The head is generally long, the face is generally oval and orthognathic, there being little or no protrusion of the jaws.

The lips are moderately full, the chin is either weakly or moderately developed." (449) In the next paragraph the Mediterranean ethnic group is described as remarkably homogeneous, yet three subgroups are recognized: basic Mediterranean, Atlanto-Mediterranean, and Irano-Afghan-Mediterranean. Montagu's criteria become even more curious when we reach the "Nordic," which is placed in quotation marks. "The so-called 'Nordic' is a mediterranean with very light skin, fair hair, and blue or grey eyes. The conventional description of the type is as follows:

The hair may be either blond, yellow, very light brown, or reddish, slightly wavy to curly in form." (450)

The third major group is referred to as the Alpine. "The typical Alpine is broad and high headed, with somewhat high curved occiput, vertical forehead, moderately to strongly developed brow-ridges, more or less roundish face, with prominent nose which has a tendency to broadness and *blobishness* at the tip." (451; italics mine)

The geographic and familial distribution of these three European subdivisions of the Caucasoid major group demonstrates that they are in no sense biological units. Individuals falling under these admittedly messy categories can and do appear anywhere in Europe and, as Montagu points out, a cross between two "Nordic" type parents might well produce characteristically "Alpine" or "Mediterranean" offspring, not to mention blends of all three. Thus, brothers and sisters in the same family might turn out to be members of three different "races" or "ethnic groups." Such results suggest that the genes responsible for the traits described are widely distributed in Europe. It is their frequencies that vary from one region to another. Thus, it is legitimate to assume that more individuals of the "Mediterranean" type will be found in southern Europe and more "Nordic" types will be found in Scandinavia. There is nothing particularly ethnic about these subdivisions, at least in the sense that this term is understood in anthropology. There is really very little reason to expect ethnic identity, which is cultural, to coincide with membership in some kind of biological group.

At this point I might be accused of stacking the cards against race as a meaningful category for the study of human diversity since I have quoted from a scientist who, in fact, does not consider this category to be terribly useful. But if we turn to the recent works of those who champion the concept of race, one by Stanley M. Garn and the other by Carleton Coon, we find that the ambiguities are by no means resolved.

In *The Living Races of Man,* Coon states:

Not every person in the world can be tapped on the shoulder and told: "you belong to such and such a race," and this fact has made some people think that there are no races at all. Ever since man first spread over the earth, interracial contacts have taken place between the populations of neighboring geographical regions, with constant genetic exchange that has produced racially intermediate, or so-called *clinal,* populations. (Coon 1965:7)

What Coon would do to remove ambiguities is to exclude these clinal populations from the main subdivisions of mankind, considering them as intermediate groups lying in zones of genetic interaction. The term "cline," however, as Coon admits elsewhere (1965: 212), refers to a gradient in the frequency of a particular trait of a single species distributed differentially in a series of populations. Different traits have different clines, so that one cannot pick out distinctive clinal populations. Some clines may be due to gene flow among a series of populations, but the term was originally employed by Huxley (1942) to refer to the action of natural selection in which favored genetic combinations change in some ordered fashion in relation to geographic gradients.

Coon sees the Caucasoid race as the most highly variable of his five major racial divisions. One might expect that his explanation of this variability would be based on the same kind of gene exchange which produces "clinal" populations. But no. Coon has most of the gene flow going the other way. Caucasoids have donated their germ plasm to both Africa and Asia, and have received little in return. In fact, he states that Negroids, or Congoids as he prefers to call them, are the result of combination between Pygmoid and Caucasoid genes. (1965:123) Although Coon sees Mongoloids as one of the purest races, with the lowest degree of variation, he gives them a fair dose of Caucasoid genes as well.

While Coon is willing to admit great variability among Caucasoids, he finds the Congoids and Mongoloids to be highly uniform in physical structure. Perhaps we reveal more about our own culture in this kind of judgment than we do about human racial variation. We know that many Americans find it difficult to tell one Chinese from another, but many Chinese have the same difficulty in respect to Americans. Chinese communist propaganda plays generally depict Americans as moustached with rather pronounced noses. For the Chinese, who tend to have small noses and less body hair than most Caucasians, these two characters submerge any differences which may exist between individuals. This fact was brought home rather forcefully to my colleague, Professor Morton Fried of the Department of Anthropology at Columbia University. Dr. Fried did his field work in China in 1949 where he lived in a small city. Another American also resided there, but lived at the extreme end away from Dr. Fried's house. Physically the two Americans were about as different as two Caucasians could be. Fried tells the following anecdote:

Shortly after World War II I was living in Ch'uhsien, a small city in the Chinese province of Anhwei. There were very few foreigners in the area, mainly missionaries who would come and go, either visiting the ancient Catholic priest who had almost forgotten his native Italian and lived in his church, hardly ever venturing out, or the younger set that dropped in on the Protestant missionary who had only recently arrived. That missionary, George Cherryhomes, was, at least in physique, the archetypal Texan, well over six feet in height, lean to the point of emaciation, fair of skin and hair. We must have made a strange contrast in the streets of Ch'uhsien, the only foreigners to be seen there with any regularity, for I am at least half a foot short of Cherryhomes' height, outweighed him by more than 50 pounds at that time, and have a ruddy complexion and brown hair.

Yet, on numerous occasions as I made my way along Chungshan Street or Great Eastern Road, I would feel a tugging at my arm and turning, would look into the face of Mr. Wang, the local postman. Beaming happily, he would bestow a fistful of letters and papers upon me and skitter away. It never failed—the mail thus delivered would be for Cherryhomes.

Finally, I confronted Mr. Wang and asked him why

this constant mistake. After all, he knew me and he should have known Cherryhomes as well. Wang studied me carefully for several minutes without speaking. Finally he shook his head sadly and said, "Wai-kuo jen tou hsiang i-yang." ("You foreigners all look alike.") (Fried, personal communication)

It was my experience in Africa among a single population located in a "typically Negroid" area of the Ivory Coast that the amount of variation in skin color, nose form, and hair color was, to my originally naive eyes, rather startling. Skin color varied from very light brown to dark brown (I have yet to see a "black" African). Nose form varied all the way from broad and flat to aquiline, and I even found occasional redheads and dusty blonds in the population. The latter trait is often missed by physical anthropologists who spend only a short time in contact with the groups they study. Blondism is considered ugly in West Africa and most people who have hair which deviates from the preferred black color simply dye it.

While I have already pointed out that the amount of physical variation within any so-called geographic race is considerably greater than most casual observers would perceive, nonetheless there is a possibility that certain external phenotypic characteristics which may be related to "racial stereotypes" do tend to be widely distributed. That is to say, such features as skin color, hair form, and facial shape may be more uniform within populations than other less obvious genetic features. If this turns out to be the case, some sort of valid explanation will be necessary. A possible explanation has been offered by Ginsburg and Laughlin (1966). These authors point out that it is useful to consider the phenotype of any organism in terms of what they call "package" and "content." "Package" refers to what is easily observable and "content" to such hidden phenotypic traits as blood groups and other biochemical properties. The authors remind us that it is possible to alter the package while keeping the contents constant and vice versa. Simply put, if we take two boxes of cereal, one of corn flakes and one of rice crispies, it is possible to substitute the contents of one for the other, producing a rice crispies box full of corn flakes and a corn flakes box full of rice crispies. It is also possible to dump both cereals into a bowl, mix them together, and put them back in their respective boxes. In the latter case the packages are different, but the contents, altered through the mixing process, are identical. This crude analogy may hold for human mating, particularly since genetic distribution in human populations is not based solely on biological phenomena but also upon sociocultural conditions, including esthetic criteria. Ginsburg and Laughlin suggest, therefore, that assortative mating in human populations, that is, mating based on the conscious or unconscious choice of certain physical characteristics, may have played an important role in shaping the genetic structure of a population. This is brought about by the social favoring of certain social phenotypes over others. Fertility could then be strongly influenced by social factors. While I think that those writers have tended to exaggerate the force and/or consequences of such a phenomenon, it may be used to explain consistency in external phenotypic characteristics within groups. Individuals may care very much about the color, hair form, body size, and shape of their mates. Only in extremely sophisticated or scientifically oriented populations would someone be concerned about the blood group of his or her mate.

But to return to the definition of race. Stanley Garn employs the following definition: "A race in man, as in any living form, is a *population,* a population of men, women and children, of fathers, mothers, and grandparents. Members of such a breeding population share a common history and a common locale. They have been exposed to common dangers, and they are the product of a common environment. For these reasons, and especially with advancing time, members of a race have a common genetic heritage." (Garn 1965:6)

While this definition is scientifically acceptable, it leads us back to the breeding population. Such a definition reduces the ambiguity of the term "race" but renders it redundant. Since the breeding population is a central concept of population genetics and evolutionary theory, it would seem advisable to use it in appropriate circumstances and either eliminate the term "race" from discourse or reserve it for something else.

Later in his book Garn takes the latter position without ever making his switch in definitions explicit. Instead he defines three types of racial subdivision which reflect a hierarchical ordering from most general to most specific. These divisions are: geographical races, local races, and micro-races. In a chapter subtitled "Race, 'race' and *race,*" Garn defines these groups. Geographical race is the largest of the three categories and encompasses the other two. It is "a geographically delimited collection of similar races. To a large extent the geographical races of mankind coincide with the major continents, and are therefore identical with *continental* races . . ." (Garn 1965:14–15)

Garn goes on to point out that geographical races

are the result of isolation produced by geographical barriers such as oceans and deserts. The ocean barrier applies most readily to the New World, which was no doubt cut off from the Asian land mass for long periods of human history, and to the Australian subcontinent which, although it may have been invaded by human beings on more than one occasion, remained much more isolated than the great land masses of Europe, Asia, and Africa. As far as deserts are concerned, their effectiveness as barriers to human travel has often been challenged. Certainly archeological evidence from the Sahara desert tells a story of settlement as well as movement across this so-called barrier both before it was a desert (only about 7,000 years ago) and after, when it was the major mercantile route between the Guinea coast and Europe. This trade continued until the middle of the fifteenth century when Europeans began to short-circuit the trans-Saharan middlemen by exchanging directly on the West Coast with coastal Africans.

Local races for Garn correspond "more nearly to the breeding populations themselves." They are largely or totally endogamous; that is, marriage occurs within groups. Typical examples of local races are Australian aborigines and Ituri forest pygmies. Garn finds this is a useful category since "clear-cut local races are largely independent evolutionary units and as such are of particular interest." This, of course, is not arguable since local races are equated with populations which are, as we have already noted, the units of evolutionary change. But the question must be asked: Are such groups really breeding populations? While it is true that they are relatively endogamous in respect to surrounding peoples, gene flow does occur, at least out of the groups in the Pygmy case and in both directions between Australian aboriginal groups.

Micro-races, Garn's third category, result from regional differences including localized selection pressures and ancient settlement patterns.

There are, to be sure, populations which share the same general living space and which are at least partial genetic isolates. However, the degree of genetic isolation between such groups is often exaggerated by the anthropologist or geneticist when the ideal rules of a culture which forbids interbreeding are taken at face value in disregard of the empirical details of everyday life. There are societies in which certain forms of marriage are forbidden, and, in fact, occur only rarely but in which there is a rather high incidence of extramarital, cross-group, sexual relationships.

It has been suggested that castes in India are totally endogamous groups and that such religious distinctions as Jewish and Christian have produced isolated populations of Jews throughout Europe and the Middle East. The endogamy of castes is related to marriage rules which involve social rather than strictly biological relationships, even though the latter may be officially disapproved. In South India, for example, there is a high incidence not only of adultery but of clandestine sexual relations between members of different castes. Hypergamy (marrying up) is a frequent phenomenon of caste systems. In addition, castes often change their status and seek new endogamous circles. While Jewish populations often differ genetically to some degree from non-Jews living near them, a comparison of gene frequencies in these groups suggests that a considerable degree of interbreeding has taken place. Thus, Jewish groups around the world differ more, one from another, than any of them do from the local non-Jewish populations among whom they live.

Such groups may be taken as separate subpopulations (when they differ significantly in gene frequencies) or as partially separate segments of a single population. The decision to lump or split such populations comes from the focus of the study concerned rather than from the application of some hard rule of genetics or a firm definition of race.

These ambiguities are the result of the potential fluidity contained in any group below the species level. The species is a relatively neat, closed unit (in spite of the many arguments current among taxonomists in regard to certain animal and plant forms that do not conform to established definitions). Members of different species usually cannot interbreed for reasons

TABLE 1

Comparative ABO(H) Frequencies from Various Jewish Groups
(After Shapiro 1960)

	Size of sample	O	A	B	AB
Holland	705	42.60	39.40	13.40	4.50
Berlin	230	42.10	41.10	11.90	4.90
Czechoslovakia	144	23.60	50.00	22.20	4.20
Poland	818	33.10	41.50	17.40	8.00
Samarkand	541	28.90	31.40	32.70	7.00
Samarkand	616	32.30	29.20	35.50	7.90
Iran	431	33.50?	32.50	25.00	9.20?
Iran	116	19.80	46.60	25.00	8.60
Morocco	642	36.90	35.90	19.90	7.30
Yemenites	1,000	56.00	26.10	16.10	1.80

TABLE 2

Comparative Physical Features among Jewish Males of Various
Groups (After Shapiro 1960)

	Mzab	Spaniol	Syrian	Iraq	East European
Eye color					
Blue (light)	11.10	6.40	10.00	1.88	14.00-20.00
Mixed	11.10	13.30		16.03	32.00-48.00
Light brown	5.60 ⎱ 80.30		25.00		
Brown	72.20 ⎰		65.00	80.19	48.00
Hair color					
Blonde					
Light brown	2.80 ⎫ 2.20		5.00	3.50	
Red brown	11.10 ⎭				
Medium brown	5.60 ⎫		⎫	⎫	
Dark brown and black	80.50 ⎭ 97.80		⎬ 95.00	⎬ 96.50	
Nasal form					
Concave	5.50	13.40		2.80	
Straight	38.90	37.30	50.00	25.23	
Undulating	11.10			1.87	
Concavo-convex	2.80			4.67	
Recto-convex	25.00				
Convex	16.90	49.30	50.00	65.42	57.00

ranging from physiological incompatibility to spatial isolation under natural conditions. When they do interbreed, most often under artificial or aberrant conditions, their offspring may be sterile. Subunits of the species almost never exhibit total breeding isolation. Once we have a unit in which interbreeding is possible, animals, being what they are, are likely to interbreed (at least on the borders of contingent populations). Any subunits established will be highly unstable and temporary, subject to rapid formation as well as extinction. Subunits of the species can arise, differentiate to some degree from equivalent units, and become extinct through reabsorption without having much effect on the persistence of the species of which they are a part. (Such units may, however, have a profound effect on the direction and tempo of evolutionary change.)

The recent history of biology has included a series of attempts to define and clarify the concept of race. In old-style biology, based on type specimens stored away in the drawers of museum collections, the concept was neat and unambiguous but had no relationship whatsoever to the realities of the living world. Under such systems races were defined as discontinuous entities, with the characteristics of one race defined absolutely in relation to all other races of the same species. One needed only to study a type specimen to know all there was to know about the group.

TABLE 3

Extent of Variation of 33 Means or Frequencies in Sub-Saharan Africa and the World Demonstrating Similar Ranges of Variation between Africa and the Rest of the World (After Hiernaux 1968)

Frequency or mean of		African range	World range	Rapport %
I^a		0.23	0.60	38
I^b	A,B,O(H) system	0.24	0.50	48
I^o		0.41	0.64	64
R_o		0.50	0.90	55
R_1		0.29	1.00	29
R_2	Rh system	0.15	0.62	24
rR^u_o		0.50	0.58	86
L^M	MN system	0.42	0.86	49
Hb^S	abnormal hemoglobins	0.20	0.20	100
Hb^C		0.14	0.14	100
Hp^1	other blood factors	0.47	0.80	59
Tf^D1		0.07	0.20	37

Frequency or mean of	African range	World range	Rapport %
Color blindness	0.05	0.12	39
	0.30	0.58	51
	0.25	0.67	37
Stature (cm)	41	41	100
Seated height technique P (cm)	7	23	30
Height of iliac spine (cm)	26	35	74
Upper arm length (cm)	16	20	82
Biacromial diameter (cm)	8	10	84
Biiliac diameter (cm)	5	6	83
Head length (mm)	21	26	82
Head width (mm)	21	27	77
Bizygomatic diameter (mm)	21	39	55
Facial height (mm)	21	26	83
Nasal height (mm)	14	14	100
Nasal width (mm)	14	15	92
Lip height (mm)	15	15	100
Cephalic index	11	19	60
Facial index	16	18	87
Nasal index	41	51	80
Cormic index technique P (Trunk to limb ratio)	3	9	33
Radiohumeral index	18	24	77

With more sophisticated field studies, and the development of population genetics, it soon became apparent that such entities simply do not exist biologically either in man or among the lower animals and plants. Few if any "subspecies" have exclusive possession of even single traits or genes. Such units came to be distinguished in terms of varying gene (allele) frequencies. That is, certain alleles were found to be quantitatively distributed in the species so that some subgroups had higher or lower frequencies of some alleles than others. Instead of neat typological units, each one separate and distinct, the biologist was left with a set of overlapping distribution curves with different means but with a great deal of overlap at the edges. This "populational" view of race or subspecies was current in biology until very recently. But the investigation of several new traits (such as blood groups, hemoglobin variants, and other blood components), easily controlled as to genetic origin and subject to mathematical manipulation, began to produce a somewhat confusing picture. Races could be easily defined or separated out if limited to a single dimension or, at most, to a small number of dimensions of loci. That is, different frequencies of specific traits could be mapped to reveal certain groupings distributed in specific geographic areas. When a large number of traits were considered together, however, the distributions frequently turned out to be discordant. That is, the geographical distribution of frequencies of one genetic trait did not match the geographical distribution of other such traits. Jean Hiernaux (1968), working with data on several hundred African populations, found that traditional classificatory schemes grouped some tribes that were widely separated on the basis of physical characteristics and separated many others which were, in fact, quite similar. All of the common taxonomic schemes failed to meet genetic criteria. More important, perhaps, is the fact that no new classificatory scheme could be substituted for the old ones. Variation tended to be continuous and showed up differently for different dimensions or trait classes.

Hiernaux's analysis, which involves the comparison and correlation of vast amounts of data, has become possible only with the advent of high-speed computers. Applied to other species the method yields similar results. Lack of concordance appears to be a general characteristic of genetic distributions in animal populations. Thus, the concept of race, which found such wide favor among previous generations of biologists, appears to have no general validity in the animal kingdom.

It must be pointed out that this remains true even when what Coon refers to as "clinal populations" are eliminated from consideration. All populations are in fact clinal with respect to some trait. And remember that the word "cline" refers to trait distributions through a series of populations. Thus, attempts to find centers of racial homogeneity now appear fruitless. It is impossible to strip away the mixed layers of overlapping populations to find a pure type underneath.

The reasons for this lack of concordance must be explained. It is due, first of all, to the fact that humans among all animals are the most unpredictably mobile and the most widely distributed of species. And wherever humans migrate they breed. The human species bears little resemblance to those territorial animals among which breeding can only take place within the home range. Unlike the salamander who, when forcibly removed from his quiet valley, will overcome almost all obstacles to return home and who will not breed while away from it, man will mate wherever his migrations take him. It must be said also that human females do not by nature display territorial prejudice.

Migration and gene flow have spread human genes around the world in myriad ways. Successive migrations, conquests, absorptions, intermarriages, alliances, and extinctions of populations have produced a constant shuffling of human genetic material.

Natural selection has also played its part in the distribution of genetic material, but its effect is often complicated and rarely well documented. As long ago as

1950, Coon, Garn, and Birdsell suggested a correlation between stature and climatic adaptation. These authors hypothesized that tall, thin individuals would have an advantage in hot climates. It was suggested that such a morphological configuration would produce rapid loss of excess heat. On the other hand, short, heavy-set individuals would be good conservers of heat produced through the normal metabolic process and should be found in cold climates. Baker, in a critique of this hypothesis, writes:

In contradiction to the original hypothesis put forth by Coon, Garn and Birdsell, experimental work has thus far failed to show that a high SA/W ratio provides any great physiological advantage to hot-desert dwellers. Beyond lowering water requirements because of a small body size, a high ratio has no appreciable effect on man's desert heat tolerance. Actually this could have been anticipated from a detailed knowledge of human physiology. The human body depends primarily upon the cooling derived from sweat evaporation for maintaining thermal homeostasis in a hot desert. The hot dry air of the desert has enormous evaporative power and is apparently capable of evaporating sweat much more rapidly than the human body can produce it. Since the sweat production of the active man is related more closely to his fat-free body mass than it is to his surface area, the total cooling per unit of weight would be predicted to be very similar for men of quite different surface areas.

Continuing on a theoretical basis, a high SA/W ratio may provide decided advantage to the man who must do physical exercise under hot, wet climatic conditions. When the air has a high moisture content combined with high temperatures, such as that in tropical forest regions, it no longer has the capacity to evaporate all the sweat produced by an active man. Under these conditions a significant proportion of the sweat will form water droplets and run off the man, providing no body cooling. Thus in a hot wet climate, the total surface area over which a given sweat production is spread will govern the amount of cooling derived. Since the amount of sweat produced by a man is governed by his weight of fat-free mass then, with activity held constant, increases in SA/W ratio should lower the heat strain on a man in the tropics. (Baker 1960:6–7)

Other explanations for the occurrence of particular morphological traits on the basis of natural selection have been offered in relation to skin color. Coon, Garn, and Birdsell suggested that dark skin is an adaptation to intense sunlight, which may produce severe burning with the risk of skin cancer. According to this theory, light-skinned people would tend to die of cancer before producing as many offspring as dark-skinned people. The heavy melanin content of the skin of Africans may function as a protective filter which cuts down on the amount of sunlight absorbed and thus reduces the risk of burning. Citing distribution maps of gradients or clines of skin color, Brace (1964) tends to agree with this hypothesis. However, he hypothesizes that dark skin is an original human trait congruent with the first appearance of man on the African continent and suggests that light skin resulted from the "probable mutation effect" which occurred with the relaxation of selection pressures. That is, as man moved north the selective pressure of sunlight was reduced or eliminated. Peter Post (personal communication) has evidence that skin deeply pigmented with melanin is sensitive to cold. Thus, a reduction in pigment would be an adaptation to colder climates.

Several other authors have also been highly critical of

the suggested basis for correlation between skin color and sunlight. Harold Blum (1961), a physiologist, has suggested, for example, that sunlight filtering devices in the skin are much more complicated than a simple deposition of dark pigment and that protection against skin cancer involves changes in epithelial tissue. It has also been pointed out that the correlation between sunlight and skin color has been based purely upon latitude placement, a simple north to south gradient, and that ultraviolet exposure is not completely contiguous with such a map.

In addition to this, some authors question the sunlight-pigment hypothesis because they believe that dark-skinned Africans originated not in the open grasslands but deep in the tropical forest where little sunlight penetrates. Coon is now one of these and he suggests that dark pigmentation is actually an adaptation to damp chilly conditions. He argues as follows: "In Europe fair skin and light eyes are commonest in zones of least sunlight and greatest cloud cover. On the other hand, in Negro Africa, the deepest pigmentation is also found in the most dimly lit regions. The critical difference is one of absolute—not relative—humidity in terms of water vapor pressure." (Coon 1965: 213–14) "We therefore suggest that one function of deep pigmentation in Negroes is to keep them warm . . ." (Coon 1965:233)

Coon emphasizes combined climatological maps in the analysis of selection pressures. This is an important point since morphological adaptation may well be tied to more than one climatic parameter. On the other hand, the fact that some authors who disapprove of the concept of race approach climatic adaptation in a rather naive way does not destroy the argument against race. Hiernaux (1964, 1968) is fully allied with those who consider race to be a superfluous concept in biology and he is also one of the most sophisticated when it comes to the consideration of multiple factors in human evolution. Hiernaux's conclusions run counter to Coon's suggestion that a closer look at the effect of natural selection will support the arguments for the existence of race.

As for Coon's suggestion that dark skin can be an adaptation to cold wet conditions, it can only be said that there is no experimental evidence for or against it as yet. One might wonder, however, how dark skin could help people keep warm in the shade. Archeological, linguistic, and ethnological data suggest that the ancestors of dark-skinned Africans originated on the high grasslands of West Africa and not in the tropical forest (cf. Murdock, 1959). This does not imply, by any means, that such a point of origin has been substantiated. Very little is known of the origin of recent African populations, and almost nothing is known of the distribution of ancient populations in West Africa.

The coincidental similarities between the peoples of Africa and the peoples of Melanesia (northern New Guinea coastal areas and surrounding islands) suggest parallel evolution toward darkly pigmented skin and woolly hair. These populations are superficially very similar, particularly in relation to external phenotypic traits. Since genetic analysis tends to show that they are unrelated, it would be surprising if natural selection had not played a significant role in the development of existing similarities. Parallels between the external phenotypic characteristics of certain members of these groups are so marked that physical anthropologists from an earlier period tended to group them under a single classificatory unit. (The more cautious among them classified Melanesians as "Oceanic Negroes.")

Other hypotheses have been offered relating morphological form to natural selection. Among the most interesting is that complex of traits which yields the so-called Mongoloid type: rather short stature, plump short fingers and toes, scant body hair, broad flat faces, with the typical epicanthic fold. Coon, Garn, and Birdsell suggested that these are all adaptations to cold, with the epicanthic fold acting as a built-in snow goggle protecting individuals from excessive glare and consequent snow blindness. The lack of hair on the body, and particularly on the face, is seen as an adaptation in that the reverse conditions presents a danger of the breath freezing in the beard, leading to frostbite, frequently followed by gangrene and death. The short stature and rather rotund bodies so typical of Eskimos place them at one end of the SA/W scale and suggest that mechanically they should be good heat conservers.

Hypotheses relating natural selection to certain single gene traits have also been suggested. Among the most famous and most convincing of these is the high incidence of certain abnormal hemoglobins in malarial regions of the world, particularly hemoglobin S in areas with falciparum malaria (cf. Livingstone, 1958). Falciparum malaria, it must be explained, is the most deadly form of this disease and is a particularly insidious killer of children. By removing individuals from the population before reproductive age, it can exert a powerful selective force. In the case of hemoglobin S the situation is fairly clear-cut. High frequencies of this gene are associated with high frequencies of falciparum malaria in parts of Africa. The genetic mechanism involved is interesting to both anthropologists and geneticists since it involves what is known as an *adaptive polymorphism*. This is a condition under which two or more alleles of the same gene are kept in high frequency in relation to one another. The reason for this is that the heterozygote form has a selective advantage over both types of homozygote. Individuals who are homozygous for hemoglobin S suffer from a fatal disease known as sickle cell anemia. Homozygous normal individuals do not suffer from this disease but neither are they resistant to the parasite of falciparum malaria. Thus, in malarial areas they are weeded out of the population. Heterozygotes do not display the more serious symptoms of sickle cell anemia, but their hemoglobin is altered chemically to the extent that it is incompatible with the development of the malaria parasite.

There appears to be a relationship between some other hemoglobin abnormalities and malaria. Among the best studied are hemoglobin C, found in high frequency in Africa, and similar conditions known as thalassemia major and minor found along the Mediterranean littoral and in India, as well as in Southeast Asia. The high frequencies of these traits in certain widely distributed populations, which might have been taken by another generation of physical anthropologists to indicate recent common ancestry, are now accepted by most as an indication of parallel adaptation to a rather severe selection pressure.

Several hypotheses have been offered in attempts to relate other blood group distributions and various blood substances to the process of natural selection. Many of these genetic traits exist in frequencies which suggest some kind of adaptive polymorphism. Beuttner-Janusch (1959; see also Livingstone, 1960) in a summary article on the ABO(H) system has pointed out that individuals with different ABO phenotypes show differential susceptibility to a wide range of infectious and functional disorders. Other authors, notably Mourant (1954), have suggested that antigenic substances present on the surface of red blood cells may have an effect on the susceptibility or resistance of individuals to specific infectious diseases, particularly plague and smallpox. Otten (1967), in a review article on this subject, has cast serious doubt on the assumed correlations. Still others (Kelso 1962; Kelso and Armelagos 1963) have suggested that differences in the distribution of ABO frequencies may be related to dietary factors. Although the literature on this subject is somewhat ambiguous, it is probable that environmental factors affecting blood group distributions will be found in such areas as disease, stress biology, and nutrition.

There is another important explanation for patterns of genetic variation found among widely distributed human groups. Here I refer specifically to genetic drift, a concept which has lost much of the original favor it had among geneticists and evolutionists. Genetic drift represents accidental change, that is, a change due to causes other than mutation, selection, or interbreeding. As noted in the preceding chapter, genetic drift can occur when a population divides through the migration of a segment to some new territory. If the new segment is accidentally unrepresentative in gene frequency, genetic drift has taken place. Genetic drift also occurs in situ transgenerationally, when gene frequencies are changed through such gratuitous events as floods, hurricanes, volcanoes, earthquakes, and certain types of epidemics. Fertility can also be reduced through the

operation of rules which produce differential access
to females. Such rules may be based on social rank,
marriages which inhibit pan-mixis (random mating),
celibacy for certain individuals, and so forth. Carlton
Gajdusek (1964) has pointed out that in small populations such circumstances are bound to occur frequently. In such populations the gene pool may be
changed once per generation through drift.

The effect of genetic drift has been thought by some
to be slight. Garn, for example, offers the following
summary criticism of the effectiveness of this process
in differentiation:

Drift, being purely chance, does not distinguish between
adaptive and maladaptive genes. . . . Drift could account
for some differences between adjacent populations, but not
differences that are distributed in a regular way forming
"clines." Only for perfectly neutral genes could drift operate alone to bring about major differences, and at the
present time we are increasingly skeptical of such neutral
genes . . . (Garn 1961:97)

Garn is certainly right that drift cannot be used to
account for traits which are distributed clinally. On the
other hand, he fails to note the possibility that drift,
followed by natural selection, could have profound effects on the development of genetic distance. Because
natural selection operates independently of mutation
and because evolution is an opportunistic process in
which selection can only operate on existing gene frequencies, any differences in gene frequency which
develop can be widened as selection operates differentially on different, related populations, particularly
those which no longer live in the same environment.
The human demographic pattern during most of man's

evolution was one of small semi-migratory or migratory groups engaged in hunting and gathering. Such groups probably had a maximal population size in the low hundreds and many may have been considerably smaller. Population growth in such groups produces fission. Although considerable genetic contact may continue among recently split groups, partial genetic isolation is an ultimate likelihood. This is a perfect situation for drift. The smaller the original population the more likely that fission will produce an unrepresentative division of the gene pool. In most cases the frequencies of different genes will be changed and in some cases certain alleles may be totally lost. Small size has the same effect on transgenerational changes, increasing the possibility that shifts in genetic structure will occur from one generation to the next.

Gajdusek (see also Cavalli-Sforza 1969) believes this phenomenon to be so widespread that he cautions against assuming natural selection when insular populations are found to have similar gene frequencies. He believes that such similarities are much more likely to be due to contact and interbreeding than to selection. In fact, he suggests that in most cases the action of natural selection on a series of island populations like those in the Pacific, even those whose parentage could be traced to a single ancestral group, would lead to differentiation stimulated first through the action of drift and continued through natural selection acting on altered gene pools under different environmental conditions.

It should be clear that several forces can act to drive populations apart genetically. But there also are forces which maintain genetic continuity. We must not forget that migration with concomitant interbreeding and gene flow through intermediate populations has been a major factor in human development. Culture acting as a buffer between man and his environment has also, no doubt, tended to reduce genetic differentiation. Nonetheless, we need a further explanation for the fact that the human species has maintained such extraordinary genetic homogeneity. Such an explanation has been offered by Morris Goodman (1963), an immunochemist. Goodman suggests that the development of immunochemical systems and viviparity have led to a slowing in the higher animals of the process of differentiation. While some immunochemical reactions develop only in response to parasitic invasion, many are inborn. The possibility is strong that mutations altering immunochemical structure will produce incompatibilities between mother and offspring. If such incompatible systems come into contact, antibodies will be formed by the mother. In egg-breeding animals these systems remain apart. Advantageous mutations producing more effective immunochemical systems will be selected rapidly with no adverse consequences. In placental animals, on the other hand, blood is transferred between mother and fetus. The possibility of harmful contact becomes an immediate problem. Most mammal orders have what is known as an epithelio-chorial placenta. Such a placenta acts as a filter between maternal and fetal immunochemicals so that the effects of mutations manifest in the offspring are minimized. However, the hemochorial placenta is a major specialization of primates. In this structure there is a great deal of intimacy between the chemical systems of mother and offspring, which increases the risk of an immune response. One such response is Rh incompatibility disease in man. An Rh negative mother who, through mating with an Rh positive man, produces Rh positive offspring will develop antibodies against the Rh factor in her own children. These antibodies will pass back into the child. Unless the child's blood is totally replaced through transfusion shortly after birth, a fatal outcome is probable.

The existence of the hemochorial placenta reduces the variation which can be tolerated between generations. The hemochorial placenta is found in very small mammals with a very short gestation period, and therefore little temporal contact between maternal and fetal blood, *and* in primates who, with their long gestation periods, face considerable danger from incompatible mutations. New mutations, regardless of their potential value in fighting the invasion of external parasites, are likely to be rejected by natural selection acting through the mechanism of maternal-fetal incompatibility. This, of course, raises the question of the value of the hemochorial placenta. Goodman notes that the type of central nervous system (CNS) typical of primates (large and complex brains, particularly the spectacular development of the cerebrum or neopallium) demands a mechanism which delivers an efficient oxygen supply to the developing brain. The epitheliochorial placenta is not adequate. Changes in placental structure favoring a shift to the hemochorial type would be advantageous only in those species in which gestation was short or those in which a rich O_2 supply was necessary for superior development of the CNS. But with the advent of the hemochorial placenta, differentiation through mutation would be considerably slowed. Such intimate contact between maternal and fetal systems favors homozygosity over heterozygosity. Goodman suggests that a compromise has been struck in primate evolution with the development of "late maturing systems" which do not produce their protein substances until after parturition. He uses this theory to account for the late maturity of primates, especially man. The point which concerns us here is his notion that the development of complex immunochemical systems coupled with the hemochorial placenta has slowed the process of differentiation or speciation. Man is held together as a species not only through gene flow but also by a feature specific to primate evolution.

Let us now return to the concept of race. I have noted that the most generally accepted biological definition equates race with breeding population. I have also pointed out that this produces an unnecessary redundancy of terms. It is equally important to note that such a definition makes it impossible for taxonomists to construct a category between the level of the species and the population. This is a rather bizarre situation since the concept of race has always been offered for just that purpose.

Hiernaux, a leading student of human variation, is skeptical that any successful grouping on this level can be constructed. If we are to derive races by organizing populations into like units, we must employ a type of analysis in which the variation present within sets of populations is smaller than the variation between sets. To quote Hiernaux:

To be useful . . . the range occupied by a cluster on the scale may not exceed the length of empty spaces between it and adjacent clusters. Suppose, for example, we are trying to classify things by their linear size, and that the total range runs from 10 to 70 cm., with an empty space on the scale from 40 to 45 cm. Two clusters appear, but two objects belonging to different clusters (of 39 and 50 cm. for example) may be much more alike than they are to many members of their own cluster. If size is considered a criterion of affinity, what is the validity of generalizing about short and long things?

Cluster analysis still applies to the case of more than one quantitative variable under consideration, but the eventual correlations between them have to be taken into account. For two variables, a geographical representation is still possible; a representation in space can be built for three variables; for a higher number of properties we can no more visualize the situation but we can make use, if a number of assumptions are satisfied, of efficient statistics . . . which still permits cluster analysis. Again the classification is serviceable only if clusters do appear, and if inter-cluster distances are higher than intra-cluster ones. (Hiernaux 1964:34–35)

In an examination of physical data (both metric and genetic) from a number of African societies Hiernaux (1968) was able to find few legitimate clusters. In the earlier work cited above he suggests that such a situation is not peculiar to Africa.

Following the above procedure would there emerge something resembling the classical subdivision of mankind into three main groups: Whites, Blacks, and Yellows (or whatever more sophisticated terms are used)? I doubt it. We know of so many populations which do not fit into the picture. Adding more "oids" to this three-fold primary subdivision would not improve it. The subdivision into nine geographic races (i.e. "the taxonomic unit immediately below the species") proposed by Garn (1961) is no more satisfactory: it only shifts the problem to a lower level. Just as Indians could not be classed with the Black or White races in the ternary system numerous populations are unclassifiable in a nine fold subdivision because they are peripheral to several geographical races. It seems highly probable to me that the more races we create the more unclassifiable populations there will be at fewer and fewer levels of differences, until we should reach a state of subdivision close to enumeration of all existing populations, i.e., the units to be classified. (Hiernaux 1964:37–38)

The reader will remember that Carleton Coon attempts to avoid the problem of overlapping populations by disqualifying them. Coon believes that by doing away with his so-called clinal populations he will be left with relatively pure centers, each characteristic of a particular race. The task is an impossible one because the mechanisms which produce distance—mutation, selection, and drift—and those which produce proximity—gene flow, interbreeding, and sometimes also selection—have acted in different ways in different places and times. Some traits vary continuously; others, in discrete clusters. Those traits for which continuous variation maps can be drawn do not all follow the same geographic lines so that no two frequency distribution maps are likely to be identical. These non-overlapping maps plus the distribution of other traits in a nonclinal fashion create a jumble which renders an objective classification of races impossible.

This does not mean that genetic distance does not vary with geographic distance. Certainly it does, or at least did prior to the last 100 years. But there are no modal points in the world which can be plotted as the loci of maximum genetic distance between, let us say, three races.

Coon counters these arguments by saying that each race has its own local clines, but this again can only be

demonstrated through the artificial elimination of intermediate populations, the very populations which con-
nect local clines of one so-called race to the clines of
another so-called race. These "clinal populations" are
the units through which genes have flowed, and they
often occur within geographic zones which reflect intermediate degrees of environmental variation. Coon's
classificatory technique eliminates those populations
which are likely to hold the key to the dynamic aspects
of genetic differentiation. Thus, his proposal not only
does not work but is scientifically uninteresting, for the
point of classification and taxonomy in science is to
create units which tell us something about process and
allow us to make predictions about variables under
changed conditions.

At this point I realize that the reader may throw up
his hands and say! "All well and good, I understand the
problems involved in defining these categories but my
eyes tell me that there are races. I can identify individuals from at least three races and classify them with
their brethren." The layman might even wish to respond that he is more willing to trust his intuition and
his eyesight than the abstract ramblings of scientists.
The scientist appears to have created an Alice in
Wonderland situation in which things are not what they
seem to be.

There are several misconceptions involved—misconceptions which produce a funny-mirror distortion of
reality. The first is the assumption that external phenotypic traits are somehow more powerful or more useful
as taxonomic devices than genotypic or phenotypic
traits which can be analyzed only under laboratory
conditions. Such a bias must be done away with. In the
past it has led animal taxonomists up some very strange

paths, since parallel evolution often produces amazing similarities between distant species and local environmental selection often produces wide separations between closely related species. Who would guess, for example, that the guinea pig and the camel are closely related? Taxonomy must be based on genetic distance. The more genes used the safer the classification. No one set of genes is better than any other set in this respect.

Second, the amount of variation present within any subdivision of man may be submerged under selective perception and biased sampling. Most black people in America received *part* of their genetic heritage from a large section of West Africa. They do not in any way represent a good sample of individuals from the entire African continent south of the Sahara desert. In addition, of course, American blacks are a mixed group with a considerable contribution of European genes.

Third, the criteria used by the average person to categorize individuals as "Negroes" are based on social rather than genetic identity. It is easy to identify an American Negro sociologically, but extremely difficult to identify all sociologically defined Negroes on the basis of biological criteria.

If we were to build a racial classification on the basis of skin color and some other phenotypic traits such as hair form, eye color, and face shape, we would get a set of maps differing from a set of maps built around one of several other phenotypic traits. There is no way of determining which of these maps is better or more correct. In fact, none are "correct" or all are "correct." The only impossible situation is that one is correct while the others are in some way false. These maps reflect different aspects of human variation and not race. It is for this reason that such physical anthropologists as Livingstone (1964) and Brace (1964) would abandon the study of race, substituting the study of clines.

On the other hand, certain eminent biologists, Mayr and Dobzhansky among them, have clung to the term "race." I think it is fair to say that for them race is an open concept which can be salvaged for certain types of work. It is not some kind of immutable grouping but rather a research tool which must be redefined each time it is used in relation to some specific research question. Washburn (1963) has stated the case well:

Race, then, is a useful concept only if one is concerned with the kind of anatomical, genetic, and structural differences which were in time past important in the origin of races. Race in human thinking is a very minor concept . . .

If a classification is to have a purpose we may look backward to the explanation of the differences between people—structural, anatomical, physiological differences— and then the concept of race is useful, but it is useful under no other circumstances as far as I can see. (Washburn 1963:252)

Thus, for those biologists and physical anthropologists who view race as an experimental tool, the reality of race is nothing more than the reality of a scientific construct which may be changed according to the particular topic under investigation. Thus, race cannot be a fixed unit, immutable in time and space. Any lines which are drawn around some set of populations are going to be artificial because the degree of gene flow or lack of such gene flow which will be used to set the absolute limits around the "race" will be arbitrary.

What this means in terms of actual experimentation is that: (1) race may be defined as a breeding population (though it has been pointed out that this is both unnecessary and confusing); (2) one may combine several breeding populations among which gene flow is known to occur or have occurred and compare them to another such set and, in the process, refer to both sets as specific races; (3) one may compare even a larger set, and so on, until one reaches the level of the entire human species. For Hiernaux even this type of definition, restricted as it is to experimental conditions, is useless. He feels that all such artificial classifications are arbitrary and counterproductive.

But the fact is that no nonarbitrary general classification of mankind is available, and what we know of the migratory habits of man, and of the extent to which population mixture took place, altogether explains why no systematic subdivision of races is possible, and eliminates the hope that a general phylogeny-reflecting classification could be constructed. Human evolution did not take the form of a growing tree, at least, not in recent times. The general picture is not one of isolated groups differentiating in circumscribed areas. Mixture occurred many times in many places between the most various populations brought into contact by human mobility. The tendency toward high adaptive specialization was balanced again and again by migration, and by man's power to transform his environment. Even if we could reconstruct the intricate succession of mixtures that contributed to each living population, the final picture would look like a reticulum more than a tree, and a reticulum defies dichotomizing subdivision. (Hiernaux 1964:42)

When I decided to write this book, I leaned toward the attitude of Dobzhansky, Mayr, and Washburn that the concept of race could be used in a restricted way to produce significant scientific results, particularly in relation to certain historical questions. My feelings were based on a dichotomy which exists within the entire field of anthropology, that is, a separation between those who are concerned with processes of human behavior, including change and development, and those who are concerned with what is known in the field as ethnohistory, the careful reconstruction of specific historical sequences for specific groups who, because they are nonliterate, have no written history of their own. Now, while these two types of research are not totally independent and while researchers in both fields hope eventually to make contributions which reflect on both subjects, the methods and frequently the results of these two pursuits are quite distinct. As far as physical anthropology is concerned, the study of clines is ideal for the investigation of the dynamic process of adaptation. Such analysis is bound to yield more information about process than some type of taxonomic device. On the other hand, I felt that for pure history, for the unraveling of the biological relationships between one set of human populations and another, the concept of race might be useful—useful, let it be understood, as an operational concept to be redefined each time according to a specific set of experimental conditions. Such a study, however, would not be one specifically limited to physical anthropology. As a historical project it would use the techniques of physical anthropology as well as the techniques of other branches of anthropology including linguistics, archeology, and ethnology. Evidence gathered from these branches could be used to reconstruct the degree of past and present ethnic similarity between groups or to unravel the physical and ethnic origins of specific peoples.

We might, for example, ask the question: have Pygmies been completely isolated historically from other groups genetically or have they had a considerable amount of both cultural and genetic contact with surrounding peoples? We might also ask whether the various "pygmoid" groups found in distant parts of the world (in Africa, South Asia, and the Pacific) are remnants of a single original population or whether they are groups which have developed similar sets of external phenotypic characteristics, independently, most likely as the result of parallel selective pressures. (See Boyd 1963.) There are several similar problems that have been raised by students of human biology and ethnohistory. Anthropologists have, for example, long been interested in the question of Basque identity in Europe. This is a population with a rather high frequency of Rh negative and blood group A. The identity of the Lapps, an ethnic group found in northern Europe, has also been queried (Boyd 1963). Such groups do not represent single breeding populations, nor are they separate species; therefore, some sort of intermediate category like race might be useful to distinguish them from other semibiological units. But this makes sense only if we can demonstrate some kind of genetic discontinuity between these groups and other groups. What holds these units together, of course, is their ethnic identity, and any biological unity which exists for them is the result of social rules concerned with such identity. Rules which reinforce ingroup marriage (endogamy), although they may often be violated, do produce genetic isolation. Thus, Jewish populations found in different parts of the world are genetically distinct from one another in large part because of gene flow with surrounding peoples, but they are also distinct from the local groups because of the partial success of social barriers erected by Jews and by those among whom they live. In this sense culture tends to preserve biological differences. Under such circumstances, however, it might be more appropriate if the term "ethnic group" were used to delimit these populations rather than the term "race." I am inclined to think that this is precisely what Ashley Montagu had in mind when he chose to substitute "ethnic group" for "race" on the level of local populations. My quarrel with Montagu is that he has added an ambiguity to an already ambiguous situation by using the term "ethnic group" in situations where there are either clear cultural differences between genetically similar populations or, conversely, ethnic similarities among genetically distinct peoples.

If we restrict the use of the term "ethnic group" to its traditonal meaning in ethnology and sociology, we shall find, I think, that it is possible to carry on biological research using a culturally defined unit as the basis of comparison. In carrying out such research, however, we must not assume *a priori* that some unit which is defined in a culturally realistic way as an ethnic group is also a biologically distinct unit.

For these reasons, I am completely in accord with those, including Montagu, who feel that the concept of race is of absolutely no value in the study of human variation and adaptation. I also lean toward those whose position is that the concept of race is a fuzzy one even when it is "carefully" defined and related only to historical questions. The use of the term "race" leads to infinite confusion.

Chapter 4

❀

THE FOSSIL BACKGROUND AND

THE ORIGIN OF RACES

FOUR MAJOR fossil groups are closely identified with
the evolution of modern man. These are: the *Australo-
pithecinae,* found mostly in East and South Africa,
associated in many cases with crude tools, and dating
from about 3.5 million to about 700,000 or 800,000
years ago; the *erectus* group, found originally in Asia,
but now known to be distributed in Africa and Europe,
dating from about 450,000 years ago and associated
with tool industries which show a marked improvement
in technique over the previous types; the *Neanderthaloid* populations, found mostly in Eastern and Western
Europe but also in the Middle East and Africa, dating
from about 70,000 years ago, with a tool complex
known as Mousterian, involving both core tools (a core
of stone chipped and shaped) and flake tools (re-
touched flakes of rock, usually flint, struck off a core);
and the *Cro Magnon* populations of Europe which are
early representatives of modern man.

Morphologically the Australopithecines are the most
primitive type, with a relatively small brain case (but,
if one adjusts for size differences, exceeding modern
and fossil apes in cranial capacity), and rather pro-
truding faces. Their limb and pelvis bones suggest bi-

pedalism and fairly erect posture. The erectus line has
a brain capacity closer to modern man than to the apes
and is distinguished particularly by large brow ridges
which project over the eye sockets and a low, sloping
forehead. The Neanderthaloids are highly variable.
Skulls range in structure from quite similar to modern
man to those with large brow ridges, low sloping fore-
heads, and no chins. The Neanderthals were equal and
sometimes superior to modern man in cranial capacity.
While there is some ambiguity about the place of Neanderthals in human evolution, most taxonomists today
would include them as a subgroup of *Homo sapiens:*
Homo sapiens neanderthalensis.

Human paleontologists have been primarily Europeans, but fossils have been searched for in many areas
of the world where geological conditions yield promising
strata. Nonetheless, the national origin of paleontologists, particularly in the early phases of the discipline,
appears to have had an effect on both the areal distribution of collections and the interpretations of these in
relationship to evolutionary history. Today important
gaps in the fossil record are being filled and a fuller
understanding of their place in man's evolution on a
worldwide basis is emerging.

One of the current problems in physical anthropology is the sorting out of the relationship between various types of Neanderthal and *Homo sapiens sapiens.*
It was originally felt that the Neanderthals were an
aberrant form, somehow cut off from the main line of
evolution. The so-called classic Neanderthal type
(long, sloping skull vault, heavy brow ridges, and no
chin) appeared to postdate more modern "progressive"
forms. It was suggested, therefore, that these fossils
represent some sort of side line, an evolutionary dead

end. Until recently, classic Neanderthal finds were limited in distribution to Western Europe, particularly France and Germany. The more "modern" types were distributed over a much wider geographical range and dated from an earlier period. On the basis of these distributions Clark Howell (1952) suggested that the classic form represented an isolated population which had been cut off by ice fields during the last glacial period. Once genetic isolation had been demonstrated, both genetic drift and natural selection could be called into play to explain the observed morphological type.

LeGros Clark (1964) suggested that the *erectus* line eventually gave way to separate lines, one leading to the classic Neanderthals and the other to modern man. This hypothesis rests on sapient-like fossils predating classic Neanderthals. Several skulls from Europe and England were brought forward to buttress this argument. (1) *Steinheim,* which, while possessing large brow ridges and a rather prognathous face, was said to be sapient in the occipital region at the back of the skull. (2) *Swanscombe,* from England, consisting only of the occipital (at the back of the skull) and the two parietal bones (at the upper sides above the temporals joining with the occipital); this fossil appeared to some to have a structure quite similar to *Homo sapiens,* but it lacks a face. (3) *Fontéchevade* man, consisting of parts of two skulls, but in total only a skull cap and part of the frontal bone; the face again is missing. An analysis of this find by Vallois (cited by LeGros Clark 1964:71) placed it firmly with *Homo sapiens.* Since it was found below Mousterian levels, which are associated with classic Neanderthal forms, it would appear to precede this group in time. Two other supposed *sapient* fossils, Galley Hill and Piltdown man, both from England, turned out to be frauds.

Brace and Montagu (1965) have referred to theories which place the rise of modern *Homo sapiens* before that of classic Neanderthals and which suggest a sudden disappearance of the latter, as hominid catastrophism. These authors opt for a more gradual pattern of progressive development through the known major lines. They place the Neanderthals in the species *sapiens*, and reject the assumption that the classic Neanderthals are a special population. Although Brace and Montagu have been the most recent authors to state this position, they credit two earlier physical anthropologists, Ales Hrdlicka and Franz Weidenreich, with the same theory. Coon also, I think quite correctly, has minimized the morphological differences between Neanderthal man and *Homo sapiens sapiens*, although as we shall see below he suggests an early origin for *Homo sapiens* in Europe and uses Steinheim, Swanscombe, and Fontéchevade as evidence.

Recently a whole series of Neanderthal finds have come to light from Central and Eastern Europe, as well as from the Near East, which display a high degree of within-population variation, ranging from classic to progressive forms. The existence of classic types outside of Howell's restricted zone tends to invalidate his argument while the wide range of variation suggests that Steinheim and other early forms may well fit into a Neanderthal phase. In a recent article Jelinek (1969) summarizes what is known of Neanderthal populations from Central and Eastern Europe and concludes:

1. The Neanderthal finds . . . display pronounced morphological variability, even at a single site (or, one might

say, within a single population). Associated with this variability is a wide variation in cultural inventory.

2. These Neanderthal finds display, to various degrees and in various frequencies, many of the characteristics that we find fully developed and universal in *Homo sapiens sapiens*. (Jelinek 1969:491)

Such variation is really not surprising, and it can be expected to emerge in some degree in other fossil groups when and if our samples approach the sample size of currently known Neanderthals.

Europe is rich in those strata in which the Neanderthaloids are found. Such strata exist, if less commonly, in other parts of the world as well, but to date most of the work concentrated on this stage of human evolution has been in Europe. The skewed sample of Neanderthaloids plus a rather chauvinistic attitude displayed by many paleontologists toward their own finds has led to the widely held assumption that modern man appeared earlier in Europe than elsewhere. It takes only a small logical leap to go beyond this conclusion to the assumption that these earlier forms were the precursors of modern Caucasoids. This view of evolution has recently been systematized and expanded by Carleton Coon in *The Origin of Races* (1962) and *The Living Races of Man* (1965).

The task of these books is to trace back as far as possible the fossil origins of five human subdivisions which Coon is convinced represent separate lines of evolution toward modern *Homo sapiens*. Once established, these five lines (Congoid, Caucasoid, Capoid [Bushmen], Mongoloid, Australoid) are hypothesized to be temporally continuous and relatively separate spatially. Although each line changes through time, each maintains biological integrity. According to Coon, each of these five lines passed their evolutionary history in much the same way. All were destined to evolve from the early hominid types into modern *Homo sapiens* through the intermediate step of the pithecanthropines (erectus). All were to be successful in attaining the sapient grade, but some were to reach it before others.

Races become differentiated from one another through the process of specific adaptation in which populations are modified to fit local environmental demands. The species as a whole, on the other hand, evolves as a unit when adaptations of a general sort spread from population to population through gene flow and then gain rapidly in frequency through natural selection which favors them in all environments. Thus, for Coon, the epicanthic fold is a specialized adaptation limited to one type of environment, while the development of a better brain has general selective value. As long as gene flow occurs, a species can evolve as a whole, but at the same time subunits can become differentiated from one another. So far so good, but Coon sees the different races of man crossing the various evolutionary levels at different times. The crucial threshold is that between the erectus stage and modern *Homo sapiens*. It occurs early in Mongoloids and Caucasians, late in Negroes or Congoids, as he prefers to call black Africans, and latest in the Australian aborigine. The assumption here must be that most, if not all, of the generalized positive mutations occurred in one, or at most two, of the five lines of evolution, and that they got to the other lines primarily through gene flow.

Most of Coon's critics find this five-line theory untenable. They cannot envisage a parallel evolution of five subspecies toward the same product. Let it be said, first of all, that parallel evolution can occur, particularly when environmental restrictions on development are quite rigid and when the evolving types come from a common stock. Parallel evolution occurs even in unrelated species when they are subject to the same highly restrictive environmental conditions. Thus many desert dwelling animals of only distant common lineage have developed a series of similar (i.e. parallel) physiological and behavioral adaptations to low humidity and a short water supply. Such adaptations involve internal water conservation through accommodations in kidney structure. Desert plants from widely divergent lineages also display certain parallel developments, all of which contribute to efficient water conservation. These parallels exist, however, as a specific set of adaptations to a specific set of conditions. Such animals and plants are divergent from one another in other morphological and physiological features because each species has its particular niche within the larger desert environment. They are different also (and this is very important) because they have different genetic histories. What are possible adaptations for some species are impossible for others because of the equipment they brought to the desert environment in the first place. No biologist expects species with parallel adaptations of this sort to be capable of interbreeding. They are too different genetically. Their similarities in structure are usually controlled by different sets of genes.

Coon's parallel evolution is of a different sort. The five units evolve toward the same general end, through a considerable time span, and maintain their interfertility. Hasty critics were quick to attack Coon for this. Coon's model, as it has been offered thus far, is perfectly reasonable from the point of view of evolutionary biology. The five units begin as one species, perhaps one "race"; they spread out and diverge, but they are held together as a single species by a considerable amount of gene flow. Centrifugal forces are balanced by the centripetal force of gene flow which acts not only to keep the species together but to pull lagging populations along with the general evolutionary trend. This part of Coon's theory is perfectly logical and reasonable, but it is probably wrong. Wrong because the human species does not stay put to the degree that some other animal species do, and wrong because human genes flow faster and in greater numbers than Coon would have it. For Coon's theory to work, you need just enough gene flow, but not too much, and you need geographically static populations. The mobility of prehistoric man was limited, to be sure; means of transport were restricted to foot travel and human populations were hemmed in by environmental barriers. At the same time, however, hunting populations (all men were hunters until approximately 10,000 years ago) are mobile. They must follow the game wherever it goes. Over the course of several hundred thousand years, game animals have been pushed hither and yon by environmental variations, including glaciers and progressively wetter or drier conditions, and possibly by the effect of hunting itself. Man the hunter has been forced to follow the game. In addition, hunting populations are bound to be rather small (although not as small as many modern hunting groups which are frequently refugees in rather barren environments). Such groups are perfect vessels for the operation of genetic drift. Rapid evolution (and the emergence of *Homo sapiens* was indeed rapid) is most likely to occur in a series of small populations connected intermittently through gene flow. Such populations act as evolutionary laboratories. In such a situation no single line can be the source of major adaptive change.

The other problem in this part of Coon's theory is the assumption that there have always been five lines of human evolution and that these lines have a historical depth and continuity which carry them back to the near beginnings of man as a species. Coon bolsters his argument for these time lags in physical adaptation by calling forth cultural evidence. If all races had a common origin, he asks, how could some people, like the Australian aborigines, still live in a manner comparable to that of Europeans 100,000 years ago? Coon responds by claiming that the ancestors of these groups and of the Europeans parted company in remote antiquity. Otherwise Australian culture would have to have regressed rapidly, a process for which there is no evidence. Coon also argues for the antiquity of separate lines on the basis of linguistic evidence. If the existing races had been rooted in one population only a few thousand years ago, linguistic divergence today could not be so pronounced. No one need argue that human divergence is a recent phenomenon to counter Coon's theory. What he does here is to set up a straw man. There is no doubt that as the human species spread out over the face of the earth, and archeological evidence shows this must have occurred early in human history, groups diverged culturally, linguistically, and physically. What is called into question is only the immutability of Coon's five groups and the subsidiary assumption that certain lines were more advanced than others. One must remember also that there *are* documented cases of rapid cultural loss with a return to technological levels below a particular set of adaptations. Such was the case for the Bushmen of South Africa who were hounded out of their natural habitat by more culturally advanced Bantu populations. Refugee populations are frequently associated with primitive technology which obscures their former cultural status.

There is no reason to assume correlations between gene flow and the flow of culture. Peoples accept genes without accepting ideas just as they may accept ideas without accepting new genes.

Coon also feels that biological differentiation in man is just too great to be of recent origin. Yet Coon once agreed that Neanderthal man would pass for an average New Yorker if he were properly dressed and found riding on the subway. Coon tends to manipulate degrees of difference depending upon his argument. At times they are very great (human races); at other times they are small (Neanderthal populations versus modern man). If we look at a species considerably younger than man, namely domestic dogs, we find a truly amazing degree of variation. Certainly the Pygmy is closer to the Englishman than the Chihuahua is to the Great Dane. True, the differentiation in dogs took place rapidly through artificial selection imposed by man, but assortative mating may have had a powerful effect on human differentiation. Add the effects of natural selection and drift to this and we need not call up great antiquity to account for the range of existing human types.

Coon argues that some contemporary races preserve a high frequency of archaic traits which can be traced back directly to the erectus stage. This is taken as direct evidence of late emergence into the sapient form. The conclusion is faulty for a host of reasons. First of all, if selective pressures (in this case technological levels) have been such that archaic traits like large jaws and big teeth tend to be preserved, then no amount of time would eliminate them from a population. Second, since there is a good deal of physical variation within all populations, drift alone could rapidly shift the average type of physical feature. This shift could be in the direction of "progressive" or "more sapient" features or toward what Coon would label more archaic or erectus features. Furthermore, there is no guarantee that the same gene complexes that give us "erectus" features in some modern populations are the same gene complexes that produced analogous traits in *Homo erectus*. Thus the use of so-called archaic traits to work out the racial history of populations of the same species is an unsound technique. The only way, in fact, to establish such evidence, if it were to exist (and this I doubt), would be through some kind of genetic analysis in which specific genes were implicated. Considering present technology, this is an unlikely proposition for human genetics.

As we have seen, Coon posits that human evolution has passed through a series of grades: *Australopithecus, erectus,* and *sapiens.* The movement of one "race" over such a grade comes through the accumulation of "advanced" genes either through mutation or gene flow and selective retention. Some of Coon's critics have suggested that his gene flow model cannot work because he has genes from one species, *erectus* for example, flowing into what are already *sapient* populations. This is, I think, another unfair criticism. While paleontologists refer to *Australopithecus, erectus,* and *sapiens* as separate species, Coon uses these terms to refer to various grades which have no absolute boundaries. If we had all the fossil evidence, we would be hard pressed to decide where *erectus* left off and *sapiens* began. The apparent differences in kind in this case are due to the incompleteness of the fossil record. Certainly gene flow among human populations has contributed to the overall evolution of the human species. Unfortunately Coon goes out of his way in both books to make the gene flow one way and to temporally separate the grade crossings for each of his five races. It is these additions to the evolutionary model which make Coon's model untenable.

In *The Origin of Races,* Coon traces what is known of fossil man from Africa, Europe, Asia, and the Pacific area. Fossils found in Europe are taken *a priori* as Caucasoid; those found in Asia as Mongoloid, etc. He tells us that the principal differences between Negritos (Pacific Pygmies), Oceanic Negroids, and Australoids are those of body size and hair form. Thus, these three peoples evolved from a common local ancestor. That these groups differ only in size and hair form will come as a rude surprise to most physical anthropologists, but Coon seems to feel that differences based on single genes such as blood groups are of little significance (*at least in this geographic area and at least in this book*). He then turns to the fossil record to build the sequence which will yield the modern Australoid. Such a sequence takes us through the erectus series found in Java, through Solo man, to Wadjak man which is given *Homo sapiens* rating in brain size but is flat-faced and broad-nosed. The jawbone has a chin (which Coon admits may or not be a diagnostic feature of *Homo sapiens*) and is massive. Other skulls—all *Homo sapiens* (Keilor, Talgai, and Cohuna)—tie Australoids into a series of Pleistocene populations in Java. The fact that the aborigine skull is massive in many of its features is taken as evidence that the Australian is still in the process of "sloughing off" a series of genetic traits which link him with erectus. C. L. Brace of the University of Michigan has offered the more sensible hypothesis that the kind of jaw we see in Australoids is the result of selective retention of a masticatory apparatus which is fitted for chewing undercooked tough meat. The retention or re-evolution of such a trait in no way suggests that Australoids are behind other sapiens in other features of evolution or that they passed an evolutionary grade later than other populations.

In his discussion of Mongoloids Coon lists seventeen features which modern Mongoloids have in common with Sinanthropus, the Asiatic representative of the erectus grade. These traits are taken from Franz Weidenrich's model of racial evolution which includes parallel development dependent upon a high degree of gene flow. Many of the traits which Coon lists are related to the chewing apparatus and are therefore implicated with the type of cultural adaptation discussed by Brace. Such traits may have a genetic base, but it is likely that some jaw and muscle development which leaves its marks on bone is the result of individual adaptation in which the body responds to a particular set of conditions. Such adaptations are not genetic. On the other hand, Coon lists the shovel-shaped incisor which is certainly genetic in origin. Brace has pointed out that this trait is common not only in Asian fossils but in many other fossil populations. It may persist among Asiatics in higher frequency, but it cannot be used as a diagnostic of fossil origin from a specific geographic area. Other traits such as Inca bones (extra bones sutured between the parietals and the occipital) are indeed a feature of Asiatic skulls. However, they are found in highest frequency among South American Indians who would appear by Coon's own arguments to be less closely related to the Asiatic erectus than are Chinese, for Indians are Asiatics who migrated relatively late in history to a new continent.

Though Coon makes no mention of it, many of the listed traits are, in fact, found in European and African fossils and some even in certain living non-Asiatic populations. No attempt is made to place these seventeen traits in the context of overall morphological pattern. Thus, Coon does not provide us with any way to determine the general importance of these seventeen traits in the determination of genetic (better, morphological) continuity among the world's populations distributed temporally and geographically.

In dealing with Caucasians in *The Origin of Races,* Coon suggests that the lack of *erectus* fossils in Europe points to an origin on some other continent, perhaps western Asia. Coon reads the archeological evidence to say that the original Caucasian populations had spread over western Asia, Africa, and Europe by the Middle Pleistocene. In the Upper Pleistocene, however, cultural differentiation took place, with the divergence of tool types in Africa. Skeletons found in Palestine, Lebanon, Iraq, and Uzbekistan are all typed as Caucasian while the contemporary African material is classed as racially different. The placement of early Caucasoid peoples in Western Asia gives them a genetic advantage, for they lie in the path of gene flow from all the Old World continental land masses. The most likely populations to have contributed their genes to Caucasians were the Australoids in India, the

Asians, the Capoid populations of North Africa, and perhaps even Congoids through contact in Southern Arabia.

The opportunity of Mongoloids to benefit from gene flow is compared to that of Caucasians. Isolation is taken by Coon as an explanation for their "extreme racial peculiarities." Europeans, on the other hand, were in a position to accept genes directly from the three other races, process them through natural selection (climatological and cultural), and pass the benefits of selection back to peripheral populations.

Thus for Coon the Mongoloid owes his sapient nature to his own mutations, but the European–West Asiatics were in a position to capitalize favorably on advantageous mutations from more than one racial group. The gift of sapient status could then be passed on to less fortunate populations (presumably Congoids and Capoids). Such mixture makes the Caucasoids the "least pure" of all human races, but this is an advantage rather than a disadvantage. This point of view, however, is directly contradicted in *The Living Races of Man,* where we are told that while the Caucasians are highly variable most gene flow was apparently out of the European area.

Much has been written about the influence of Negroid infiltration into Mediterranean countries and about Mongolian genetic penetration in Eastern and Central Europe. Both have been exaggerated. . . . The Moors, that is the Arabs and Berbers, occupied much of Spain and Portugal for seven centuries and Arabs also held Sicily for a while. Arabs and Berbers are themselves Caucasoids, but they brought a number of Negro slaves with them. (Coon 1965: 66)

I find it extremely difficult to imagine a tremendously long period of contact between a whole series of populations, with a wide range of ethnic identity, existing around a navigable lake which was fully exploited, giving rise to a one-way, or predominantly one-way, flow of genes. Logic demands the contrary hypothesis of two-way gene flow and differentiation resulting from drift and natural selection. Circum-Mediterranean populations all share many genes. They are what Coon refers to as "clinal populations."

Turning to fossil material, Coon delineates a series for the Caucasian line going back to the Mauer mandible, which has characteristics similar to both African and Australoid groups but is divergent from the Mongoloids. The turning point for Caucasians is the Steinheim skull dated to the Mindel-Riss interglacial 250,-000 B.P. or 110,000 years younger than Sinanthropus (the Chinese erectus). Coon stresses that Steinheim's cranial capacity (1150–1175 cc) does not differentiate it from the Javanese and Chinese erectus line but that other features of the skull show an advance over the erectus grade. This is particularly noticeable in the occiput (the base of the skull), which is rounded. The forehead is low but quite steep; the mastoids are small (which by the way is an apelike characteristic); and the side walls are parallel.

The next skull in the series is from Swanscombe in England. Its cranial capacity of 1275 to 1325 cc puts it in the modern European female range, and the measurements for breadth and height closely resemble modern Caucasoids. Coon goes on to say: "There has been a great deal of speculation about Swanscombe's face, but because Steinheim has a face, and because the threshold between Homo erectus and Homo sapiens lies in the brain and not in the face it is unnecessary." (Coon 1962:495)

The involuted reasoning of this quote is a milestone in physical anthropological analysis. My translation would read as follows: It is the brain size which determines the grade and not the face, but the face of Steinheim is sapient and not erectus and therefore it is likely that the face of Swanscombe is also sapient! This analysis stands in strange contrast to that given to the Rhodesian skull, which is classed by Coon as erectus and by many, if not most, other anthropologists as Neanderthaloid.

Fontéchevade and Swanscombe are also offered as evidence for early sapient emergence in Europe with the additional assumption that these early populations were Caucasoid. LeGros Clark agrees with Coon in respect to early sapient emergence but makes no mention of racial identity. Even if the case for rapid evolution of *Homo sapiens* in Europe were proved, it does not follow that these populations were Caucasian. As we have seen, recent evidence tends to support the argument most recently put forward by Brace and Montagu that Steinheim, Swanscombe, and Fontéchevade fall into the widely divergent Neanderthal grade.

In *The Origin of Races* Coon devotes a chapter to Africa, the home of the Capoid and Congoid subspecies. He admits that the evidence for Congoid origins is so scarce as to be difficult of interpretation but he nonetheless builds a sequence by marrying existing fossil material to speculation about the effect of natural selection.

Most physical anthropologists would dare hazard a guess about the race of a skull, but they would much rather have a complete skeleton and would be considerably happier with a fairly large sample skeletal population from which they could draw average measurements. Yet Coon is able to extract the following information from the incomplete skull (it lacks the lower jaw) known as Rhodesian man, a fossil put out of the sapient species by Coon himself:

In the frontal index of facial flatness and in the simotic index (reflecting the archings of the nasal bones at their root), the Rhodesian skull falls within the Caucasoid range, and in the third or rhinal index of facial flatness it resembles the ancient Egyptians and approaches the means of published series of living Negroes. In the fourth or premaxillar index of facial flatness it leaves all modern populations far behind it. In other words, this face is Caucasoid in its upper portion, Congoid in the middle and virtually pongid below. On the whole the face is mostly Negro . . .

The tibia, which was found in the bottom of the cave with the skull, resembles that of a modern Negro in all essential details. (626)

In addition, we are told that Asselar man, another fossil from West Africa, was a Negro from the neck down!

Coon needs to have the Capoids (or Bushman) originate in North Africa and so a myth is cited as evidence. The Riffians are said to have a vivid image of their predecessors:

They had the ability to transform themselves: a *thamza* [female] could turn into a bewitchingly beautiful Berber damsel, and an amziv [male] into a Negro. *Obviously then, they were, in their natural forms, neither Caucasoid nor Negro.* (601–602; final italics mine)

And then Coon refers back to fossil material: "If the Ternefine-Tangier folk were not the ancestors of the Bushmen, they were a sixth subspecies that uniquely died without modern descendants, and the Bushmen would have no discernible ancestors." (602)

Turning to the effects of natural selection, Coon suggests (589) that the Negro originated in the savannah lands at the edge of the tropical forest. Black skin is offered as an adaptation to possible deleterious effects of strong sunlight. But in *The Living Races of Man* Coon offers an entirely different theory already noted above: "We suggest that one function of deep pigmentation in Negroes is to keep them warm." (Coon 1965:233) The Negro homeland is now the deep forest away from the direct rays of the sun. The forest environment is damp, and in the rainy season the combination of relatively low temperature and high humidity can produce chills. Dark skin which would absorb more heat from the hearth than light skin provides an advantage even in an environment in which the sun's rays rarely penetrate to the forest floor.

In *The Origin of Races* Coon has the Pygmies developing out of the Negro with a penetration of the tropical forest. In the second book (*The Living Races of Man*) Pygmies emerge as the original black population of Africa. On page 100 we are told that the Pygmies were probably the original inhabitants of West Africa and on page 105: "We can generalize that in every measurable or observable character known the Pygmies stand at one extreme the African Caucasoids at another and the Negroes in between." On page 123 an analysis of crania indicates that Negroes gravitate between Mediterraneans, Caucasoids, and Pygmies: "This evidence suggests that the Negroes are not a primary subspecies but rather a product of mixture between invading Caucasoids and Pygmies."

But Coon has already told us that mixtures do not produce races, that most of the gene flow was the other way (from Pygmies to Negroes), and he never includes Pygmies as a sixth racial group. Not only do the two books contradict each other but they tend to garble Coon's whole theory of racial origins. Since they represent, in Coon's own words, two parts of a single work, it is difficult to ascribe differences between them to progress in the author's thinking. Certainly Coon does not indicate that the contradictions represent a major change in his original theory.

The reader who accepts the dictum of *The Origin of Races* will be surprised to find out that "today the indigenous population of Africa is mostly clinal. In the Sudan and East Africa, Caucasoids shade into Negroids; and telltale pockets of partly Capoid peoples survive in the Sahara and along its northern fringes." (Coon 1965:84) So a whole continent is relegated to clinal status. If this is the case, we have to eliminate Negroes from our discussion for Coon has already informed us that clinal populations don't count. But if the cline extends downward into Africa surely it extends upward into Europe as well. This would leave us with only two real races: the Mongoloid and the Australoid!

In *The Origin of Races* Coon has genes flowing into Europe; in *The Living Races of Man* he has them flow out of Europe. Perhaps Coon wants us to read these contradictions out of existence by adjusting to temporal changes in gene flow. Thus, early in fossil development there was flow into Europe, and later, a flow out of Europe into Africa; but even if we were to grant this possibility, which is highly dubious (why the change in direction?), the whole argument destroys his theory of races as entities, at least for Negroes and Caucasoids.

Coon admits that the fossil evidence which he calls forth to support his claim for late emergence of the sapiens grade in Africa is poor. The Chellean-3 skull from Olduvai gorge, which is classified by most physical anthropologists (including Coon) as *Homo erectus* and dated contemporaneously with both Heidelberg and Sinanthropus, is offered as a possible ancestor for both Caucasoids and Congoids. Coon rests his argument on the Rhodesian material, not only for the divergence of the Congoid line, but for the late evolution of this line from the erectus to the sapiens grade. As I have already indicated, most physical anthropologists place Rhodesian man in the Neanderthaloid group, above erectus on the phylogenetic scale of hominid evolution, and none to my knowledge would care to attribute any racial affinity to it. Coon's interpretation of this single skull is at best highly dubious.

A skull from Cape Flats, South Africa, is said to follow the Rhodesian pattern in which Caucasoid and Negro features are combined. This and another fossil, the Border Cave skull, classified as "Australoid" by South African physical anthropologists, are taken as Congoid by Coon who sees South Africa as a cul-de-sac for the remains of early Congoids. He asserts (1962:633): "Whether or not a local race of Negroes evolved in South Africa before the ancestors of the Bushmen arrived has little to do with the origin of the Congoids as subspecies . . ." Coon states flatly that the Negro originated in West Africa, an area from which there is "not a single scrap of evidence."

The oldest "Negro fossil is that of Asselar man, found near Timbuktu." This, I must remind the reader, is the post-pleistocene specimen wholly Negro from the neck down. Coon cites other skeletal material but it dates from such a recent period that it adds little substance to his argument.

In *The Origin of Races* Coon suggests that the Negro, who was originally adapted to savannah living, developed adaptations to the forest environment through intermixture with Pygmy groups. This is rank speculation, there being no evidence whatsoever for the borrowing of a forest genotype on the part of Africans from some other group, nor for the origin of Negroes through the admixture of Pygmies and Caucasoids (Coon's second theory of Negro origins). And then Coon is faced with the self-stated problem: Who are the Pygmies? His solution is another flight of fancy. He agrees with Gusinde and with Gates that Pygmies are descendants of old, pre-Hamitic, pre-Capoid populations of the African savannahs who were driven into the forest by drought.

Coon's first theory is that an original proto-Negro, proto-Pygmy group crossed with true Pygmies to produce the modern Negro (1962:655–66). It must be stressed again that we know virtually nothing of the origin and development of recent African populations. We also know nothing of early differentiations. If (as is possible) Black African populations emerged late, this would reflect only one case of the kind of divergence that can take place within a polytypic species over and over again through the operation of the genetic processes described in this book. A possible late emergence of the Black African, however, has nothing to do with Coon's hypothetical late crossing of the line

between erectus and sapiens. For racial or, better, populational differentiation can take place within a species
at any stage of development. In *The Living Races of
Man* Coon suggests that Negro origins can be tied to
a cross between Caucasians and Pygmies late in the
biological history of man. The arguments, therefore,
that Negroes are somehow inferior to Caucasoids,
which have been employed by some of Coon's *readers,*
make no sense on any grounds.

I have one other major criticism of Coon's work.
The photographs of members of different human populations presented in *The Living Races of Man* have
little scientific validity for comparative purposes because they are in no way controlled for differences in
age, sex, dress, hair, style, expression or even angle
of photograph. Furthermore, they lend support to typological arguments because they are open to interpretation as "type specimens."

We are shown, for example, four Basque males
(plates 77–78 a, b, c) to show "a wide range of facial
features"; three Tuaregs (plates 107 a, b, c)—"One of
the expected lean and aquiline type" and "two other
Tuaregs who look like ordinary Berbers." But plate
109, we are told, is a Berber with "Bushmanlike features." Such photos tell us nothing of real variation of
physical type within such groups. One might reply that
Coon has done his best in assembling 128 plates cov-
ering a wide range of peoples. But since these pictures
have no scientific purpose, why include them at all?
The captions are also misleading, to say the least, for
many refer to the ethnic or linguistic identity and give
not the slightest clue to genetic affinity. Thus under
circumpolar peoples we find (3a) Zyrian, "The Zyrians
are Finnish people hunting and herding reindeer in the

forests of northern European Russia" and (3b) a
Vogul, "The Voguls are Ugrian-speaking peoples of
the Obi River country who live principally by fishing."
Plate 11c is a "Siberian woman with a Ukrainian
father and an Eskimo mother." Coon tells us that "she
looks completely European." Because of her haircut
and her general facial structure, she looks like a Japanese to me, but no matter. Plate 33 is "a Japanese
nobleman of aristocratic facial type." There is no mention in the text of differential physical types among
Japanese social classes. In plate 38 we are shown a
"Tibetan lama with curly hair"; but we are never told
the frequency of this trait among Tibetan lamas or any
other Tibetans. Plate 43 represents three Nagas from
northeasternmost India and adjacent parts of Burma.
"The Nagas resemble American Indians in appearance." To me the man in the picture does, the women
don't—but *which* American Indians, and how representative are these Nagas?

The most misleading, even scandalous, set of pictures, however, appears in *The Origin of Races,* Plate
32, which shows an Australian aborigine woman and
a Chinese man with the following caption: "The Alpha
and Omega of *Homo sapiens:* An Australian aboriginal
woman with a cranial capacity of under 1,000 cc.
(Topsy, a Tiwi); and a Chinese sage with a brain nearly
twice that size (Dr. Li Chi, the renowned archeologist
and director of Academia Sinica)." The inference is
that Li Chi's intelligence, of which we have no measure,
is vastly superior to that of Topsy and that the difference is due to brain size. Topsy could be a genius.
Correlations, within the normal range, between brain
size and intelligence do not exist. Females usually
have smaller brains than men not because they are less

intelligent but because they are smaller on the average than men. Finally, although Coon gives us an estimation, we don't really know the cranial capacity of either Topsy or Professor Li Chi. The picture with its accompanying caption tells us more about Dr. Coon's preconceived notions than about either Dr. Li Chi, or Topsy, or the mechanism of intelligence in the human species. It certainly has nothing to do with the origin of races except that for Coon the Australian aborigine is the result of a late advance into the *sapiens* grade of human evolution.

Many of Coon's critics, in their attempt to be fair, have stated that *The Origin of Races* is an excellent compendium of fossil man, and so it is. But the book was not written as a catalog and cannot be judged as such. The old-fashioned biology which it represents and the errors, particularly in genetic and evolutionary theory, which it propagates, serve, in my opinion, to retard scientific progress. I find *The Living Races of Man* to be as inaccurate in substance as *The Origin of Races,* although I think Coon is justified in taking exception to the rather simple bio-environmental hypotheses which others have offered to account for the distribution of particular physical characteristics.

❀

Chapter 5

OF MICE AND MEN: BEHAVIORAL

GENETICS AND HUMAN VARIATION

THE JACKSON LABORATORY in Bar Harbor, Maine, has been raising pure strain mice for several decades. Many of these animals, the result of years of incestuous matings, have been bred for the specific purpose of advancing cancer research. Recently a new generation of geneticists interested in the rapidly expanding field of behavioral genetics began to examine these mice for strain-specific behavioral characteristics.

All mice have a natural repertoire of "mouselike" behavior including species-specific learning patterns. The question was: did the Bar Harbor mice display strain-specific differences in behavior? Remember, these mice were not bred for behavioral differences. The experimental evidence provided an affirmative answer; in most cases strain-specific behaviors did emerge. These ranged from abnormal responses to certain stimulation, particularly noise, to peculiar locomotor patterns associated with progressive degeneration of the central nervous system. Most interesting, perhaps, was the fact that among all the Bar Harbor species there was one which preferred alcohol to water! These mice did not learn to drink. When naive mice were offered a choice between 10 percent alcohol and water in identical bottles, they demonstrated a high preference for the alcohol. These animals known as C57/Bl/crgl, a black variant of the C57 strain (hence the Bl), were the only mice of six strains tested which showed this preference. An analysis of the trait revealed that a single gene was responsible.

Evolutionists have long realized that a major trend in the phylogenetic development of animals has been the increasing complexity of the central nervous system associated with a concomitant increase in behavioral complexity. Such evolutionary trends have enabled higher organisms to absorb more information from the environment, and to process it for mobilization into adaptive behavioral responses. The ability to organize, store, and utilize information and to act in response to particular stimuli has taken two forms: preprogrammed responses and learning. The preprogrammed response (which is frequently referred to as instinct) is triggered by some environmental or internal stimulus. I prefer the term "preprogrammed response" because the concept of instinct tends to suggest that innate responses are automatic and invariable. This is frequently not the case. Experience, particularly during the period of maturation, may alter preprogrammed responses considerably. In fact, while learning and preprogrammed responses may appear to be different, they intermesh to a considerable degree. Under natural conditions, as we shall see below, a wide range of behavior in infrahuman species appears to be automatic and invariable. The interaction between experience and automatic behavior emerges most clearly, however, under laboratory conditions. The essential difference between these two behavioral modes is that with the programmed response there is a high probability that under conditions of specific stimulus a naive animal will respond in a patterned and predictable way. Learned responses, on the other hand, generally involve a period of trial and error.

Under certain conditions a single trial may be sufficient to alter behavior. Mice normally run from the light into the dark; from an open to an enclosed space. This is a preprogrammed response. If a naive mouse is presented with a flight-stimulating situation which drives it from the open through a small hole onto an enclosed platform containing a shock grid, a single shock will be sufficient to alter its behavior. The mouse learns that if it goes through the hole it will experience pain.

Learning also involves a capacity to respond to stimuli with appropriate behavior. Capacities for learning are inherited, but not what is to be learned. This provides an animal with considerable adaptive flexibility. While such capacities vary in content and complexity, the ability to learn extends downward in the animal kingdom, perhaps as low as single-celled organisms. In man the capacity has been highly developed, including the ability to learn language and culture, a totally nongenetic set of traditional behavioral patterns.

Increases in the complexity of the central nervous system (CNS) allow for a greater amount of learning but also for increased complexity of preprogrammed responses. Among lower forms of life, protozoa, for example, such responses are very simple and generalized. These organisms may increase random movement in response to stimulation and thus eventually escape a stimulus field. There is no directional element in this response. In more complex organisms potential activity may include movement toward or away from specific stimuli. In multicellular organisms, such as the arthropoda, such relatively simple responses may persist but we find more complex behaviors as well; web spinning in spiders is a good example. Complex responses of this type include nest building and courting behavior widespread in fish, reptiles, and birds, and nurturant behavior found in birds and mammals. Toward the top of the phylogenetic scale (as measured temporally and in terms of CNS development) precoded responses become less and less important. This is particularly true of the primates and, above all, man. An expanded capacity for learning and the ability to transmit this learning through language provide the human species with a flexible form of adaptation, much more useful in the long run than stereotyped responses, particularly since the human species occupies so many ecological niches.

There is now solid evidence that the preprogrammed forms of behavior are influenced by genes in the same way as morphological traits. It is just as evident that certain perceptual mechanisms are species specific and that capacities for learning are also inherited. It is very important that these facts not be confused with the frequent assumption that because a gene or genes which affect behavior exist, the behavior must manifest itself in some fixed way. The behavioral geneticist Ginsburg has said: "All aspects of an organism may be thought of as one hundred percent genetic, but not one hundred percent determined." Just as the penetrance of a somatic gene and its final expression in the phenotype may be influenced by other genes and the environment, a gene which affects behavior may be modified as the result of gene-environment interaction. When behavioral scientists investigate the inheritance of certain behaviors, they are dealing with phenotypes—the outcomes of these interactions. The geneticist deals with phenotypes and genotypes. He assumes that the analysis of genetic mechanisms will depend upon a full understanding of the genotype while a full understanding of the behavior will depend upon the analysis of the historical interaction process between a particular set of genes and a specific set of environmental conditions. There are many routes to the same phenotype, and much genotypic variation is buried in phenotypic expression.

A major problem inherent in the investigation of the genetics of behavior is the definition of behavioral units and the translation of these definitions into operational (experimental) terms. What is aggression, for example? Is it to be measured in terms of thresholds of response to increased degrees of stimulus, or in the frequency of attacks, or on the basis of persistence of attack, or all of these? Once aggression is defined operationally for one species, will that definition suffice for other species? Is the unit of behavior under investigation single, or multiple with different dimensions controlled by different genetic units? For example: is timidity-aggression a single dimension on a possible scale of behavior or are there separate behaviors which yield degrees of timidity and degrees of aggression? If the latter, timidity and aggression could vary independently, while they could not if the former held true. In sum, behavior is more difficult to define and measure than such morphological traits as eye color or skin color.

In addition, one must be extremely careful to define what specific effect the gene or genes are having on a particular system of behavior. Is the effect due to central nervous system function? Or is the endocrine system involved as is so often the case? Could the CNS and the endocrine system be working together? On the other hand, could genetically based difference in behavior be due to simple differences in perceptual abilities? Some strains of mice are deaf. Others have poorer vision than normal wild strains of mice. Often inbred strains are weaker in general than wild strains. Behavioral differences therefore may be due to some general physical condition. In addition to this, similar highly characteristic behaviors may be due to different genes or different combinations of genes and environmental stimulation.

Thus it is not adequate to state simply that a certain area of behavior is controlled or affected by genes. It is important to get at the root of this structure in relation to morphological, physiological, and eventually biochemical functioning.

The difficulties raised here are illustrated in part by a long series of experiments which began in the 1930s in the laboratory of Professor Tryon of the University of California at Berkeley. Tryon's work demonstrated genetic control over maze running ability in rats. He developed two strains of rats separated by their ability to perform in a learning maze. Good learners were separated from poor learners and each set was inbred for several generations. Such a process is analogous to natural selection, but the selecting conditions are controlled by the experimenter and the matings are artificially close. Such selective inbreeding has been carried out by animal husbandmen for centuries in attempts to shape behavior and morphology to the requirements of the breeder. This process has resulted in better egg layers, better milk-producing cows, higher ratios of meat to bone in cattle, more aggressive fighting cocks, good hunting dogs, aggressive watchdogs, and so on.

Tryon's experiment was quite successful. After just a few generations, strains of rats emerged which came to be known as the S_1 brights and the S_3 dulls. The S_1 rats learned their maze running problem at a significantly higher rate than either randomly selected rats or S_3 rats, and the S_3 rats performed significantly less well than their S_1 cousins or ordinary rats.

It seemed for a time as if Tryon had selectively developed strains of intelligent and stupid rats; that it was, in fact, possible to select for intelligence in animals. The situation is, however, somewhat more complicated than Tryon realized at the time. Several years later Searle demonstrated that the results could be reversed if the type of maze were changed. What had happened was that Tryon had been subjecting his animals to an enclosed maze in which nonvisual cues were primary to learning. When the rats were placed in an open maze in which vision played an important role in the learning process, the S_3 rats performed better than the S_1 animals. These results can be interpreted in two ways. One can say, for example, that intelligence is a highly complex phenomenon made up of differences in perceptual ability and skills and what Tryon had done was to select for one type of intelligence rather than another. Or one can say that what Tryon had actually done was to select animals with particularly good and particularly poor kinesthetic sense modalities and that it was this difference in perceptive ability alone which biased the result of the experiment. But these two interpretations are not really that different. The only measures we have of intelligence at the present time are based upon performance, and in many cases it is difficult to objectify the reasons for performance differences even when they occur in regular fashion. I shall have much more to say about the problems inherent in intelligence testing and intelligence as a concept in Chapter 7.

Recently a team of scientists at Berkeley took another look at Tryon rats. This time they were interested in the differential effects of environment. S_3 rats were separated into two groups and raised under contrasting environmental conditions: one enriched; the other deprived. The enriched environment contained devices calculated to stimulate physical and perceptual development. The deprived environment left the rats under relatively constant conditions of low stimulation. After several weeks in these contrasting conditions, rats were run in maze learning experiments. Striking differences in performance were manifest between these two groups of genetically identical rats. Those animals which had experienced the enriched environment performed significantly better than those which had experienced the deprived environment. Unwilling to accept a simple correlation between behavior and environmental stimulation these scientists, under the leadership of Krech, Rosensweig, and Bennet, began a careful analysis of the brain morphology and chemistry of the two sets of animals. It was found that there were differences in brain weight, morphology, and chemical activity between the two groups. My colleague Professor Ralph Holloway found preliminary evidence that microscopic differences exist in the density of dendritic branching (nerve connections) in the brains of these rats: a higher degree of dendritic branching was found in a small sample of brain sections coming from the enriched group. It was also found that the level of chemical activity in the brains of the enriched rats was higher than in the deprived group.

While facile extrapolations from rat to human behavior should be avoided, one cannot help noting the similarity of performance between deprived rats and deprived human children. Current thinking in childhood education leans toward the theory that lack of stimulation in early childhood has a stunting effect on intellectual development. The principle behind such programs as Head Start is that children should experience early intellectual stimulation. If the theory is correct, then one must conclude that the hereditary element in intelligence is only one important factor in development. Americans have been remarkably successful in tapping the hereditary potential for height through good nutrition. Intellectual potential could probably be brought to its hereditary limits through careful and thoughtful education. One could better accept inevitable hereditary differences in intellectual capacity if all children were given the chance to manifest their total potential.

There is another environmental aspect of brain function which must not be overlooked, however, if we are to improve intellectual output. Cravioto, Delicardie, and Birch (1966), in a study of the effects of nutrition on brain function, offer strong evidence that early episodes of malnutrition, particularly protein deficit, can have long-range, sometimes permanent, effects upon intellectual capacity. The research was carried out in Guatemala among peasant populations where malnutrition is common. After eliminating hereditary factors, they found an extremely high correlation between nutritional status and height. Small and tall children were then compared in a series of psychological tests which suggested that both reversible and irreversible brain damage may result from malnutrition.

Several recent experiments on crosses between two pure strains of the same species reveal that hybrids may perform better than either of the parental types. This is consistent with genetic theory in which a phenomenon known as hybrid vigor may occur. John H. Bruell, for example, experimented with exploratory behavior in pure strain mice and hybrid mice.

Mice placed in a strange environment behave as if exploring it. To obtain a measure of such activity we placed mice individually in a four compartment maze. As the mouse moved from one compartment of the maze to another, it interrupted a light beam and activated a photo-relay and counter. The exploration scores for an animal consisted of the total count registered in ten minutes of testing. (Bruell 1965:129)

The experiment was highly complex. Five out of thirty-one groups scored below the parental average and only one group below the low scoring parent. On the other hand, "twenty-one groups scored higher than the higher scoring inbred parent. Thus, overall, we can speak of heterotic inheritance of exploratory behavior. As a group, hybrid mice tended to explore more than their inbred parents." (Bruell 1965:132)

This result is very significant because explorative behavior is a major feature of adaptation in animals. The same kind of result in hybrid mice has been achieved for other areas of behavior by Harry Winston of the University of Pennsylvania. It must be stressed, however, that these phenomena must not be taken as a simple effect of genes on intelligence or some other aspect of behavior. It has long been known in animal biology that inbred strains tend to be less viable than wild types. The reason for this is quite simple. An animal that is homozygous on many loci is exposed to the danger that lethal and sublethal and other types of debilitating recessive genes present in the gene pool of the original population will be unmasked. Thus a restoration of heterozygosity may increase viability and behavioral vitality through a decrease in the amount of genetic load manifested in individual animals.

So far most behavioral genetic studies have been limited to "lower" animals. A good deal of the work has concentrated on such abnormal behavior as autogenic seizures in mice (the convulsive response of several strains of mice to loud, sudden noises); the alcohol-imbibing qualities of certain mouse strains; learning and exploratory behavior of several animals in various rat populations; sexual and space orientational behavior in fruit flies; and a series of extremely interesting experiments with dogs relating genetics to such behavioral parameters as aggression, timidity, tamability, and so on.

It must be noted, too, and this is extremely important, that most of the behavioral genetic studies we have to date have been carried out on pure strains in the laboratory. The geneticist has tested for genetic factors by holding environment constant and manipulating controlled genetic units. This is a useful approach because the geneticist is most interested in the genotypic and ultimately the chemical foundations of behavior. But pure strains are very different from wild populations. Even the dog strains tested by Scott and Fuller (1965) were a far cry from wild species (dog breeds have been under the selective control of dog breeders for hundreds of years). Historically, I think it is fair to say that the earliest genetic manipulation by man was carried out on dogs, the first domestic species. Generations of selective matings have produced wide morphological and behavioral differences in dog subspecies.

Very few experiments have been performed, to my knowledge, on the behavior of wild strains brought into the laboratory where the genetics can be tested against a constant environment. (Work has been done on some wild rodent species and also with birds.) There have been, however, a rather large number of experiments and observations on wild and domestic animals which attempt to separate elements of learning from precoded behavior. These studies are usually performed by animal psychologists and ethnologists who tend to sidestep questions concerning specific genes and their actions. Kilham and Klopfer, for example, noted an experiment (by Wecker) in habitat selection in deer mice.

A laboratory reared strain of mice derived from a field dwelling race showed no strong preference for a field over a woods habitat. By rearing some young in each of these two habitats, Wecker established that an experience in the field early in life could produce a preference for that habitat. However, a similar experience in a wood did not produce a corresponding preference for a forest habitat. Here, too, the preference for fields was immanent in the sense that a particular experience was required in order for a response (choice of field over forest) to be elicited; a different experience could not elicit a different preference, however. Presumably, deer mice derived from the forest dwelling races could only be "trained" to prefer the wooded habitat, but this experiment has yet to be performed. (Kilham and Klopfer 1968:23)

Klopfer, in the same paper, citing his own experiments with chicks, demonstrated that

Naive chicks, whether of a yellow or a black variety, show no consistent tendency to approach other chicks of the same variety; when communally reared in the light with chicks of their own variety they do develop a preference for their own kind. However, when reared with chicks of the alien variety, no consistent preferences appear. Preferences for own kind are immanent but require activation by a particular experience. Programming of perceptual systems so that an experience is prerequisite to a response, but with only certain kinds of experiences or responses being possible, seems to be an important feature of organisms. (Kilham and Klopfer 1968:24)

Work done on the so-called imprinting phenomenon in birds by such pioneers in ethnology as Tinbergen has also demonstrated that the following response of a newly hatched duck will generalize to any moving, noisemaking animals or object. But this will occur only within a specific time period, after which the response that under normal circumstances produces a fixed behavior will fail to occur at all.

The problems which might arise from loose structuring in the behavioral evolution of many species are eliminated by the fact that under wild conditions such animals as black chicks are unlikely to encounter yellow chicks early in life, and ducks are likely to see mother duck rather than some probing barnyard scientist. Tatania imprinted on Bottom in *Midsummer Night's Dream* because Puck had put a magic juice in her eyes. She fell in love with the first creature she saw on awakening. This is the only known case of imprinting in adults of any species and of humans of any age. The combination of "a stable or probable" environment with such loose genetic determination is normally powerful enough to produce an appropriate and adaptive response. But experimentally induced aberrations of these responses further enforce the notion that behavior, even genetically determined behavior, is susceptible to a wide range of environmental influences.

Investigation in human behavioral genetics which gets down to specific genetic mechanisms has been limited almost exclusively to abnormalities. The two most famous cases concern the successful analysis of the genetic element in Down's syndrome (Mongolian idiocy) and the much less successful speculation over the role of genes in schizophrenia.

Advances in laboratory fixing and staining techniques coupled with methods for arresting chromosomal division in cellular material have enabled cytogeneticists to discover the genetic basis of Down's syndrome. It consists of an extra twenty-first chromosome (twenty-one triploidy). Such aberrations appear to be responsible for the morphological and behavioral symptoms associated with this condition. Other forms of mental deficiency have been related to genetic disorders, but less is known of the cytogenetic phenomena involved. In the case of phenylketonuria, a genetically based metabolic error is responsible for early brain damage.

There is also some evidence that an extra Y or male sex chromosome (twenty-three triploidy) produces personality disturbances which may even lead to antisocial behavior. Twenty-three triploidy also produces a set of morphological characteristics including a tendency toward severe acne. While it is possible that genetically induced chemical imbalance is directly responsible for the assumed abnormal behavior, it is also possible that such behavior could occur in a high percentage of affected individuals as a result of secondary factors such as feelings of inadequacy developed in response to physical handicap.

In the case of schizophrenia the picture is even more complicated and difficult to unravel. There is a great deal of controversy in psychiatric circles over the possible role of genetics in schizophrenia. There are essentially three schools of thought although it is sometimes difficult to categorize a particular theory. The first, common among psychoanalysts, particularly Freudians, holds that schizophrenia is caused solely by social factors, particularly abnormal socialization. The second believes that schizophrenia is a genetic disease unaffected by environment. The third believes that individuals may inherit a tendency toward schizophrenia but that life stress is responsible for its manifestation. Such arguments sound like the nature-nurture controversy limited to a specific case. I would bet on the third school, since genes for schizophrenia would be just as likely as any other genetic phenomenon to be affected by the environment. The problem is difficult, however. Schizophrenia does appear to run in families, but such data alone do not separate environmental from genetic factors. Studies of identical twins compared to two-egg twins and other siblings tend to confirm a genetic basis for the disease. Concordance figures from a number of American and some older European studies are impressive, but not all twin data are equally satisfying. In addition, psychologists and psychiatrists have some difficulty in defining the disease as a specific entity. It manifests itself in various subtypes of behavior, and the occurrence of similar symptoms in small children and adults may not be related to the same disease or cause. The fact that schizophrenia tends to occur with a high frequency (about one percent) in all populations tested suggests that, if it is genetic, the genes involved would have to confer some benefit on carriers. Biochemists have attempted to isolate chemical substances from the blood and urine of schizophrenics and their normal relatives, but so far the results have been at best equivocal.

Research in comparative human psychology has frequently revealed behavioral and perceptual differences among human populations. Spuhler and Lindzey, in a review of such research, cite the following:

Although the earliest psychological comparisons of races dealt with simple sensory processes and modes of response, there has been relatively little systematic work in this area until very recently. Indeed the decades immediately before and after 1900 probably saw more pertinent investigation of this variety than we have seen in the ensuing 50 years.

One of the earliest experimental comparisons of races was conducted by the pioneer American clinical psychologist, Lightner Witmer. . . . He compared Caucasians, American Indians, and a group of mixed African-Caucasian descent in reaction time to visual, auditory, and tactile stimuli. The American Indian subjects had the lowest average latency, followed by the African-Caucasian group, with the Caucasian subjects the slowest to react.

The well known Cambridge Anthropological Expedition to the Torres Straits included among its staff the psychologist-anthropologist Rivers and the psychologists Myers and McDougall. . . . Rivers, who directed the psychological studies, focused his own efforts upon the study of vision. He found that the non-literate subjects were generally superior in visual acuity to typical European norms but he expressed strong doubts concerning the dependability of this racial difference . . .

Myers studied auditory acuity and tone discrimination, finding that the natives of the Murray Islands tended to be generally inferior to European subjects, whereas there appeared to be no appreciable difference between the two groups in the upper limit of their hearing. A similar study concerned with olfactory discrimination and taste discrimination led to the conclusion that "we may say of the Murray men that their sense of touch is twice as delicate as that of Englishmen, while their susceptibility to pain is hardly half as great." (Spuhler and Lindzey 1967:376–77)

The authors also mention Rivers' work among the Toda of India where differences in visual and tactile ability were found.

They turn next to the Saint Louis World's Fair of 1904 where 400 subjects of various racial backgrounds were studied. Experiments with auditory acuity showed Caucasians performed better than various Asiatic populations, a finding which reversed earlier results. Clear-cut findings were also brought forward by Woodworth to show Caucasian superiority in color discrimination, lower pain thresholds in Europeans, and greater visual acuity in Asiatics.

Turning to more recent studies, Spuhler and Lindzey note:

Although there were several rather isolated British studies of racial differences in "phenomenal regression" during the 1930's, only in the past decade has there appeared any substantial evidence of interest in racial and cross-cultural differences in perception. Allport and Pettigrew compared European subjects, acculturated Zulu and nonacculturated Zulu subjects in the incidence of perception of the trapezoidal illusion. They found that under optimal conditions there appeared to be little difference in the performance of the three groups, but under marginal or nonconducive conditions the unacculturated subjects reported the illusion less often than the other groups of subjects.

As a part of a similar but much more extensive and systematic program of research Campbell and Segall have reported a study of racial and cultural differences in the incidence of various perceptual illusions. . . . This investigation is in many ways a model of how to go about making comparisons across cultures or races, particularly because of the painstaking efforts made by the investigators to distinguish between variation in the effectiveness of communication and actual differences in the psychological process under study. The research involved the collaboration of a large number of anthropologists and psychologists and spanned a substantial number of non-literate tribes as well as American subjects. The investigation reported marked group differences, . . . we find, for example, that the incidence of the Muller-Lyer illusion is almost four times as frequent among American subjects as among Bushman subjects . . .

It should be noted that none of the American investigators adopts a nativistic frame of reference in accounting for the racial differences in the frequency with which these illusions are reported. They consider the most reasonable explanation of the differences to be in terms of variation in the environment or experience of the different races. . . . This interpretation does not fit in neatly with all the data, for, among other considerations, there is the fact that young children are more susceptible to the illusion than older subjects in our culture; one might reason that, if experience in the "carpentered" world produced the illusion, the illusions should become more common with greater exposure to this environment. (Spuhler and Lindzey 1967: 379–81)

The fact that children are more susceptible to illusion than adults may, in my opinion, relate only to the guarded approach of adults to the testing situation. If this is the case, one might not find a similar guarded approach among naive subjects who have had very little experience with testers and interviews. We must not forget that Americans are highly test conscious and that this consciousness may skew results in many directions. Experience with testing may help one to score higher, but suspicion on the part of the individual tested that what he is seeing is not in fact "real" may reduce his score on illusion tests. In addition, such perceptual phenomena are frequently encountered in children's books and games. Grown-ups are likely to have had such illusions explained to them.

On the other hand, Gregory (1966, 1968) has offered evidence in support of the theory that all optical illusions are effectively caused by accommodation to what "should" be seen in an orderly world rather than by some physiological process of perception. If he is correct, and his evidence is based on excellent experimental work, then the results of many, if not all, of the illusional tests will have to be explained on the basis of environment and experience.

The studies cited by Spuhler and Lindzey do not sort out genetic and environmental factors. The groups investigated are referred to as ethnic (cultural) and racial when in fact most of them are largely ethnic. We have no genetic control over the so-called Negro or Caucasoid samples. The Torres Straits study and some of the other works cited may reveal interpopulational differences which may be due in part to genetic factors. But it must be remembered that what is being compared in these cases are populations (with specific ethnic identities as well) and not races. Spuhler and Lindzey tend to slide, as do many other authors, between a definition of race which is coterminus with population and a definition of race as a subspecies defined in quasi-genetic/quasi-geographic terms.

It must be noted that the tests under discussion concern perception and not behavior. The next step would be to relate differential perception to behavioral differences. Recent studies of cognition demonstrate that the natural world is different in different ways to different peoples. Color categories are often widely divergent among ethnic groups. The division of the spectrum into a different set of categories might influence the perception of these categories, just as differential perception could lead to the formation of different categories. The evidence so far, based on acculturation studies, learning, and the observed contrast between small perceptual differences and large category differences, favors a cultural explanation for test results which show divergences in perceptual ability. Environmental factors must have their effect on learning as well. Turnbull (1961) notes that Pygmies used to living in the forest, where the depth of their visual field is limited, have difficulty in judging distances on the open plain and often see large animals that are far away as small rather than distant.

It must be underlined that Spuhler and Lindzey make no attempt to dichotomize results into superior and inferior classes in their review and analysis of perceptual and behavioral differences. The average scores of populations will be relatively high on some tests, lower on others.

The most important point to be made about the studies cited by Lindzey and Spuhler is that they all are concerned with phenotypes rather than genotypes. The phenotypes dealt with are undoubtedly a highly complex product of interaction between genetic potential, environmental factors, and individual experience. Since the responses to these various tests in each population tested are likely to show only average differences, i.e., differences in the means of responses, it would be interesting to trace similarities and differences in family lines, tying tests into genealogies as well as group membership. While this type of testing does not control for or eliminate environmental influences, similarities and differences in family lines compared to population means should add further useful information to the analytic mill. With the exception of twin studies it is difficult, if not impossible, to separate genetic and environmental factors in human behavior.

Behavioral output is an extremely complicated process influenced by such factors as memory, motivation, interest, span of attention, emotional thresholds, ability to perceive, and ability to separate differential stimuli or to generalize and group them into appropriate categories. Sensory motor coordination is also important, as is the emotional attitude toward the task to be performed. All of these factors may be influenced by both hereditary and environmental factors. Furthermore, just as two types of computer can solve the same problem in different ways, problem-solving behavior in humans may follow different routes to the same results. What goes on inside each computer or individual will depend upon its particular structure, and the operations performed may be totally different. Thus, while no one denies that differences exist in the behavioral potential of different humans, it is extremely difficult to sort out those factors which are hereditary in nature.

When it comes to group comparisons, the difficulties are compounded. Any human population is going to display a range of behaviors, a range of thresholds, and degrees of perceptive acuity. Whether or not average scores in these several variables will differ from population to population is in each case open to question. In addition, the degree of variation within population aggregates is likely to be larger in many cases than the degree of variation between such aggregates. (This supposition rests on the work of Hiernaux and others who have already demonstrated such distributions for somatic traits.) The essential comparisons will have to be made between populations and not "races." At the present time those differences among populations which might ultimately be traced to genetic influence are difficult if not impossible to uncover. It would make sense to begin with studies of identical twins to discover the most likely genetic variables and then to extend these studies to populational units.

Logically one should expect to find neither great differences between populations nor high uniformity within any one group because the evolution of the human species has put a premium on more or less the same range of differential behavior, no matter what the environmental setting. Man is a social organism in every sense of the term, and social groups depend in part upon behavioral differences among their members. (Similarities are also important.) In addition, all human beings as individuals perform a variety of social roles the emotional and affectual content of which demands a wide range of behaviors. One can be a stern parent and a loving sibling or offspring at the same time. One can be tender to intimates and harsh to one's fellows if conditions of leadership make this kind of demand upon an individual.

Although man's genetic structure evolved in hunting

Homo sapiens neanderthalensis
(classic type from
Le Moustier   part of cheek-
bone missing from cast)

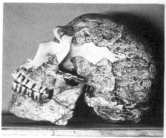

Homo sapiens neanderthalensis
(progressive type, Skull 5)
Note the chin, rounded head
and large brow ridges

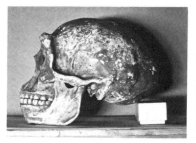

Homo erectus (*pekinensis*)

Homo rhodesiensis

Steinheim man

(All photos by Alexander
Alland, Jr., from casts
in the laboratory of
physical anthropology,
Columbia University)

FIGURE 4 *Fossil types*

Chimpanzee and *Homo sapiens* (modern man)

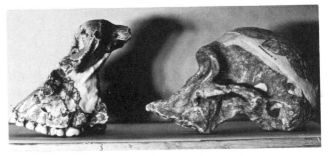

Australopithecus robustus and *Australopithecus gracilis*

Australopithecus robustus and *Australopithecus gracilis*

societies, he manages rather successfully to exist under the incredibly changed and incredibly complicated conditions of modern industrial society. Many physical anthropologists have made much of the fact that only a few generations separate us from our hunting ancestors. Since genetic change is slow in comparison with culture change, they argue that man is still basically a hunter, at least as far as his genetic potential is concerned. It is suggested that maladaptation results from this lag. What these authors tend to neglect is the rather fantastic accommodation that the human species has made to urban living with all of its artificial problems. Modern society certainly demands a greater degree of behavioral polymorphism than early society, but a wider range of talents, personalities, and intelligence are tolerated and exploited under modern conditions than could have been when man lived in small, closely knit social groups with primitive technologies. It is culture not genetics which has brought hunting populations to the verge of extinction in the world. Everything we know about the evolution of man has pointed away from a rigid behavioral genetic structure and toward a flexibility coupled with an increase in the evolution of open capacities. The evolution of capacities (Spuhler's 1959, "the capacity for culture") has led to a considerable stripping away of the preprogrammed behavior, for such preprogramming would be disjunctive with the open code system which lies at the basis of all human social systems. This leaves man to cope with environmental variables creatively. It also allows him to occupy a tremendous range of habitats.

Given man's unquestionable variation in both morphology and behavior, the question remains: Is there any correlation between physical structure and behavior? If so, are there any direct genetic links which operate to produce these correlations. Early attempts to relate somatic characteristics to behavior (like those of Lombroso who diagnosed criminal types from facial features) have fallen into disrepute. The last major effort in this direction was Sheldon's work in the 1930s and 1940s (see Sheldon, et al. 1940). Employing photos of a large sample of nude human figures, Sheldon attempted to derive an objective classificatory system of physical type based upon components of fat (endomorphy), muscle (mesomorphy), and leanness (ectomorphy). Each characteristic was scored on a seven-point scale. Sheldon also rated his subjects for behavioral characteristics and sought correlations between these and each somatotype. While there is some doubt that Sheldon's system successfully isolated three separate physical dimensions (there is evidence that ectomorphs and endomorphs lie at the extremes of the same dimension), his was a pioneer attempt to find physical markers for particular behavioral types. Sheldon's work did not include genetic analysis, and although he was convinced that his somatotypes and their correlate personalities were inherited and immutable, the data are disputable. Sheldon's research stands as a kind of crude behavioral genetics without the necessary genealogical and genetic data. Several cogent criticisms of his work have been published, and for several years his name was relatively forgotten in comparative human psychology. Recently, with the rise of behavioral genetic analysis and the new search for morphological markers of behavior, Sheldon's ideas have been revived to some extent. Today his studies point the way, but also stand as a warning against poor methodology and hasty conclusions. A good deal more caution and attention to design, methodology, and careful interpretation of experimental results are demanded by current work in this field.

In a recent paper Gardner Lindzey (1967; see also Rees 1960) has suggested that it might be worthwhile to reinvestigate morphological correlates of behavior. The problem inherent in such studies begins with experimental design. Lindzey's paper is an important step, for instead of merely accepting sets of relationships, he stresses the necessity for explaining the dynamics and cause of such relationships and suggests several alternative explanations. The first concerns the effects of a common environment on both behavior and morphology. Thus, for example, "If we accept certain psychoanalytic formulations concerning the relations between obesity and a particular type of parent-child relation—maternal overprotectiveness—and assume, moreover, that physical components are influenced by weight and diet factors, we would confidently expect some degree of association between the endomorphic and fat factor and those behavioral consequences that are believed to be associated with maternal overprotectiveness." (Lindzey 1967:229) The second explanation concerns the rather simple observation that an individual's behavior is limited or facilitated by his particular physical type. It is unlikely, for example, that an endomorph would have much success as a ditchdigger, or an ectomorph as a heavyweight boxer. On the other hand, an extremely tall individual who was well coordinated might make a good basketball player.

The third factor is a social one in which individuals of particular physical types are likely to be recruited by society on the basis of their physical type for particular behavioral roles. The fourth factor is also social but rests upon cultural stereotypes which associate particular physical characteristics with particular behaviors. The suggestion here is that social expectations might drive individuals to behave in expected ways. Thus, one might find a high correlation between red hair and temper, fatness and good humor, and so on.

The final possible factor (the most objectionable to the majority of American psychologists) is a single biological determinant of both behavior and physique. Research in this area must begin with intrapopulational studies. Correlation must be sought between different social roles, behaviors, and morphological dimensions. These must be tied to genealogical data and twin studies wherever possible. If valid correlations are established and explained satisfactorily, it might then make sense to engage in interpopulational comparisons to see if the expected relationships hold in different cultural circumstances. It is possible that when certain relatively clearcut physical characteristics are found to be widely distributed in specific populations, mean behavioral differences and the ranges of behavior might show correlated interpopulational differences. We know, for example, that Eskimos tend to be short and rather rotund with short extremities and that, at the other extreme, Nilotic peoples in East Africa tend to be tall and thin with long extremities. (It must not be forgotten, however, that even these two divergent and geographically separated populations are genetically very close.) Such extreme differences might produce consistent behavioral correlations which are based on that complex of physiological, genetic, environmental, and cultural factors which undoubtedly account for the diversity of mankind. But any conclusions which stressed genetic causality to the exclusion of other factors would probably be wrong, for genes alone do not determine phenotypes.

Ginsburg and Laughlin have recently examined the theoretical implications of behavioral genetics for studies of human populations. They begin by attacking the hermetically sealed cultural approach to behavior, pointing out that human behavior is an end product of biological and cultural factors.

The congruencies of morphology, physiology, and behavior in Eskimo adaptation to cold stress and to glare contain genetically based intradependencies within a bio-behavioral reticulum rather than interdependencies between disparate entities. In contrast, interdependence between Eskimos and their dogs is empirically based upon a practical knowledge of breeding and behavior. This in itself, and similarly with other groups of hunters, is an important part of their adaptive system. The intellectual achievements of primitive groups have been consistently underestimated and underreported; and to the extent that these constitute an integral part of their adaptive system, they should form a focus for more intensive studies. (Ginsburg and Laughlin 1966: 240–41)

The authors point out that the achievements of mankind, wherever man inhabits the earth, demonstrate the

. . . enormous plasticity and the potential for achievement inherent in populations generally. Thus, the Maya Indians of Guatemala and Mexico who generated a concept of zero, mathematics, astronomy, writing, monumental architecture, etc., developed these expressions of biobehavioral abilities on their own from a population base of primitive hunters who had entered the New World no more than 15,000 years ago. The fact that populations as remote from each other as Eskimos and Mayas can adapt to environments that are exceptionally diverse, with differences in intellectual interests, growth rates, disease patterns, nutritional regimes, population densities, mating practices, etc., and yet remain similar in many marker genes, suggests that, in fact, we know very little about the epigenetic systems and homeostatic devices shared by diverse populations that place limits on their genetic and phenotypic change and thus maintain the genetic unity of the species.

The genetically diverse populations inhabiting similar habitats, provide a natural comparison matrix in which research can be efficiently pursued. (Ginsburg and Laughlin 1966:241)

I think that Ginsburg and Laughlin's remark apparently suggesting that the concept of zero and other cultural accomplishments of the Maya are linked to genetic change could mislead the layman. If what the authors meant was that the genetic potential for such behavior is present in all populations, then the statement is perfectly acceptable and congruent with what we know of the genetic background of human behavior. If, on the other hand, they mean to suggest that certain genes facilitate specific types of invention, I think they are in error. Certainly the authors do not mean to imply that specific behavioral outputs involving particular intellectual products can be linked directly to any gene or set of genes. The wording of the sentence in question, however, leaves this reader with the uncomfortable feeling that Ginsburg and Laughlin intend more than a simple expression of nascent capabilities. The stage is set, however, for a program of research involving the analysis of the interplay between cultural and biological variables.

The authors criticize those approaches to human behavior which dismiss genetic causality. This point of view began in American anthropology with Boas, who criticized the one-sided theories of behavioral geographers and racial determinists. The latter erred by attributing differences in behavior to biological and environmental factors alone. The legacy of Boas was certainly necessary and healthy for it purged behavioral science of unscientific, basically racist theories, but it also left an unfortunate dichotomy in its wake: biology vs. culture. In place of this dichotomy Ginsburg and Laughlin substitute what they refer to as a holistic synthesis involving the already quoted view that all aspects of any organism may be 100 percent genetic but not 100 percent determined. (Not 100 percent determined meaning that the phenotypic outcome is the product of genetic and nongenetic factors.) The authors would turn anthropological research toward the underlying factors which produce phenotypes, including the genotypic background. They see the human species as a population reticulum "where genetic exchange occurs at a higher frequency in cultural-biological aggregates than between such aggregates. The adaptation of these populations has led to a cultural and biological stability such that the acceptance of new genes in an established gene pool through gene flow does not produce a highly noticeable difference in the standard 'accepted' phenotype." (Ginsburg and Laughlin 1966:267)

Thus they cite the case of American Negroes in which about 20 percent of the genes are of European "white" origin with a small admixture of genes from Indian populations. Yet they feel there is a relatively fixed Negro phenotype. As noted in Chapter 3, I wonder how much of this "fixed phenotype" is the artifact of differential perception on the part of the non-Negro observer. I would say at the very least that more objective analysis of phenotypic variation within groups of populations other than "Caucasian" be carried out before we accept these principles of lower variation in groups other than our own. This is an important question because Ginsburg and Laughlin suggest that the phenotypic stability is the product of assortative mating, a phenomenon which theoretically could go a long way in explaining nonselective (in the Darwinian sense) aberrations in expected gene frequency. I have used this concept myself to explain the possible retention of certain genetic conglomerations in apparently phenotypically similar populations which actually differ considerably from one another in "invisible" genes, i.e., those which have no noticeable effect on physical features. Assortative mating could operate among a set of culturally dominant populations as in parts of Africa, but I don't think one can make the same generalization for the American Negro, a minority group which has until recently accepted the aesthetic standards of the dominant white majority.

Although I suspect that the American Negro provides a dubious case, Ginsburg and Laughlin's concept of package vs. contents already alluded to is, I think, a useful one. The point they make about such a dichotomy is that mechanisms which tend to maintain the continuity of the package also allow for the absorption of a good deal of hidden variation. Cultural-aesthetic factors shaping the package would have little effect on "hidden genes," but they could create and maintain visible intergroup differences. Such a phenomenon would account in part for the kind of gene distributions which were employed in the static classifications of race as well as the discordances which make racial classification impossible. The point which Ginsburg and

Laughlin make about this hidden variation is that all human populations have an internal source of genetic material which can be mobilized under conditions of selection pressure. Thus, even if behavior were 100 percent genetically determined, all human groups would have more or less the same potential for successful adaptation. Ginsburg and Laughlin are convinced that the genotypic variation hidden under sets of similar packaging is of such a magnitude that the reduction of the human species to a single population of reasonable size, perhaps in the neighborhood of 100,000, would not reduce the chances of new cladogenic development. Thus, as this single population grew and spread, new phenotypic divergences would appear in connection with specific environmental demands.

Since these extensive genetic resources are present in all populations, and the limited phenotypic expression of these resources will vary along a wide set of independent dimensions, Ginsburg and Laughlin reject the idea that human differences can be sorted out in some kind of superiority-inferiority axis. Populations will respond differently to certain sets of behavioral tests, but no population is likely to have exclusive priority over the entire upper range of such tests. And, of course, behavioral measures relate to phenotypes rather than genotypes and therefore do not reflect the true range of genetic variation within any particular group.

Laughlin, who has had considerable research experience among various Eskimo groups, has suggested that those populations which use kayaks to hunt on the open sea face a stringent set of selective forces which sort out navigational abilities. These hunters often set out in fogs away from the land and spend several hours on the sea well out of sight of any landmarks. In order to return successfully, they must keep track of wind changes, changes in current, turns that they make, and, at the same time, they must concentrate on the hunt. An Eskimo hunter who performs poorly will be automatically selected out of the group. Unable to find land he will ultimately drown or starve. Since hunters are the younger mature men, their loss will have an immediate effect on gene frequencies. In this case selection pressures are extremely severe.

While the genetic explanation for Eskimo success in navigation is certainly interesting, one must not exclude out of hand an element of learning in the development of these peculiarly Eskimo abilities. When the argument is extended to other hunting groups, it loses some of its strength, particularly if it is used to account for the development of hunting skills. In most societies the hunter runs little risk. Furthermore, his kill is usually shared with the entire social group so that differences in ability tend to be canceled out through social rather than through genetic mechanisms. On the other hand, where hunting bands are small and the degree of consanguinity close, it is possible that the sharing phenomenon can keep a tight gene pool going. In this situation, differences in hunting ability might be significant for whole groups, with selection favoring those groups in which certain behavioral and physical skills prevail.

The hunting life is not, or was not, as difficult as it has often been painted in the anthropological and romantic literature. Many, if not most, contemporary hunting populations represent remnant groups which have been pushed out of favorable habitats by more advanced agricultural and urban peoples. In addition, the supply of game has been whittled down in many parts of the world through overhunting with modern weapons and by man-induced environmental changes. Marshall Sahlins (1968) has pointed out that even in recent history many hunting populations have had a relatively easy life, with plentiful game. Also, relatively low population density decreases the risk of disease in these groups and so selection pressures on them may not have been as severe as one might suspect. We must not forget, at the same time, that genetic drift may change gene frequencies in small mobile populations. If many behavioral genes are buffered, as Ginsburg and Laughlin suggest, then drift could have a strong effect on them. The existence of differences in the frequencies of "behavioral" genes, when such genes occur, could then be due to chance factors as often as to selection.

I think it is fair to say that if a genetic basis exists for behavioral differences *within* populations, then given those factors which operate on gene pools to change gene frequencies, it is probable that interpopulational differences exist as well. What is not clear, however, is what these differences mean in terms of actual behavior and what the causes of such differences might be.

Ginsburg and Laughlin suggest ways in which this process can be analyzed, and list six possible explanations for expected correlations. (1) Correlations are due to a single or common agency. The genotype controls the endocrine system. Such a system is in turn responsible for somatic and behavioral characteristics. (2) The correlation may be brought about by the operation of the same genes, but somatic and behavioral

traits are dependent upon different mechanisms. (3)
The correlation may be due to closely linked but independent genes. (4) The correlation may be due to
co-adapted mechanisms which are based upon separate
genetic units. (5) Genes may also be paired "through
independent selection for separate effects." That is,
they may go together because similarities in environment have independently selected for them. (6) Correlation may be due to chance alone.

In contrast to Lindzey (cited above) the authors do
not include the possibility that such correlations result
from social factors overriding genetic mechanisms.
Nonetheless they take this problem into consideration:

Data on the invariant occurrence of particular associations
of traits in widely different populations as against the possibility of their occurrence in different associations in diverse populations (as well as within a single, variable population) are needed to help eliminate the range over which
speculation may occur and also to identify, by means of
the differences emerging from such comparative studies,
what the traits are that may be considered to be natural
units of variation for human beings and how they are distributed in the context of reticular organization of the human species. (Ginsburg and Laughlin 1966:272)

It is to their credit that the authors offer other
methodological cautions:

Behavioral genetic analysis has demonstrated the following
facts:
(1) A measure of central tendency with respect to a
behavioral attribute of a genetically variable group provides very little useful information. The isomorphism of

the attribute being measured with the way in which the organism's behavioral capacities are organized is not to be inferred from the fact that the members within a given population can be measured in this particular respect. Individuals graded alike on such a measure may yet be genetically different with respect to the behavioral potential the test purports to measure.

(2) While it is true that behavior is complicated, and that behavioral capacities are likely to be affected by many genes, it does not follow that the genetic substratum for such capacities are unanalyzable or that particular major genes do not exert major determining influences on these capacities or potentials.

(3) Many seemingly highly correlated phenomena have no necessary causal connections so that data demonstrating correlations among physical, physiological, and behavioral attributes in particular populations must be further interpreted in order to evaluate the significance of the correlations obtained. (Ginsburg and Laughlin 1968:29–30)

The necessary separation of hereditary and cultural factors and their reintegration in a theory of biocultural interaction upset most research designs. It seems to me that the final unraveling of the problem will depend upon biochemical analyses in which a direct link between chemistry and certain behavioral potentials is established. Even here it will be necessary to demonstrate that chemical differences are genetic in origin rather than artifacts of environment or behavior. Several years ago a group of scientists were convinced that they had isolated a schizophrenia-specific substance from the urine of hospital patients which could not be found in a nonhospitalized normal population. The substance turned out to be an artifact of institutional living, particularly diet. It was subsequently also found in prison and hospital populations among individuals who had no history of mental illness.

Biochemical analysis in combination with carefully controlled genealogical studies should be added to correlational studies in which the mean gene frequencies of one population are compared with the means of another. The latter technique serves best (and none too well it must be added) as a technique of classification, while the former techniques search out the dynamics of process.

In a recent paper, *Boldness and Judgment in Behavior Genetics,* Herbert Birch (1968) provides a judicious caution. Birch stresses that what appear to be standard results of behavioral tests may be misjudged and misinterpreted. Certain specific environmental effects often exert a powerful influence on experimental outcome. Factors which appear to be genetic in origin may, for example, turn out to be related to such phenomena as the maternal effect. This results when contact with the mother produces particular physiological or behavioral characteristics in the offspring. Such phenomena may mimic genetic effects since differences follow family lines. Furthermore, Birch notes that animals selected for high performance in one set of environmental circumstances may perform quite poorly under another set of conditions. Thus, the genetic potential for a specific set of behaviors is also a potential for a level of performance under a stated set of conditions. When the conditions change, the results may also change. He also reminds us that age-determined differences may occur in the interaction of organisms with their environment.

Birch finds the genetics in behavioral genetics

"sometimes sound, but almost always the behavioral analysis is terribly poor."

What one is given is the end product of learning in a given maze, or the N score differences in discrimination, or mean differences between groups in scores on intelligence tests, etc., without any serious efforts to determine what characteristics of the organism as a response system are involved in the mastery of the presented problem. Animals tested for "intelligence" in one set of perceptual modalities may do very poorly while when tested for the same type of cognitive ability in another set of perceptual modalities may do extremely well. Such findings apply not only to Tryon rats, but to rat learning in general. (Birch 1968:54)

Hirsch (1968) criticizes what he calls a "typological" reification of behavior. This is the assumption that various constructs such as motivation, intelligence, aggression, anxiety, etc., are realities which can be measured confidently across populations and even across species with the same techniques. Not only may the tests be structured in such a way that results are incomparable but assumed similarities of behavior may not actually be related to the same set of phenomena.

Behavioral genetic research is beset with a series of problems which must be overcome if such research is to be meaningful:

1. Heredity and environment must be separated for experimental purposes.

2. The additive effect of heredity and environment must be understood since phenotypes result from gene-environment interaction.

3. Reification, in which definitions of behavior are taken as universal realities, must be avoided.

4. Factors producing behavioral differences must be

separated. Are they based on performance abilities
(linked to CNS or endocrine system), perceptual differences, learning, or on a combination of all these
factors?

5. The specific effects the testing situation has on
results must be determined.

In sum, genetic effects on behavioral *potential* are
surely real phenomena. It is highly probable as well
that differences in the frequencies of genes related to
behavior occur among different populations. But such
a conclusion provides no clues as to how these differences are manifested or even if they are manifested at
all. It certainly cannot be used to argue that differences in behavior between populations must be based
on this or that genetic factor. Nor can it be used to
support what are essentially irresponsible statements
concerning a biological basis for behavioral differences
which occur among those large population aggregates
which have been referred to as races. Nor can we expect that populations exhibiting the same mean values
for certain behavioral or perceptual tests are necessarily
equivalent in genetic structure. Groups which are similar in external phenotypic characteristics may be highly
divergent in "behavioral genes" and groups which are
different in such phenotypic characteristics may be similar in the frequencies of these genes.

Chapter 6

❀

THE SOCIOLOGY OF RACE

SOME YEARS ago during my undergraduate days I spent a summer with an archeological team on the Mandan, Arikira, Hidatsa reservation in North Dakota. Although my workday was spent exclusively with a shovel, my spare time was devoted to a study of local reservation customs. During the course of the summer I was fortunate to meet several members of the three tribes and to attend a series of ceremonials. I also became acquainted with the local Indian Service schoolteacher and his family. The teacher, an Amerindian, was married to a white woman. They had two children, a boy of five and a girl of three. I remember my first meeting with them. After a few minutes of conversation, the children toddled up and the mother introduced them as "my two little Potawatamies." The children looked up at her and in unison responded, "We're half white." The mother blushed a little and said, "My children don't understand how they can be half white and all Indian at the same time." This anecdote illustrates the difference between biological myth and social reality in racial identity. The mother responded in terms of a standard social definition; her children responded in a naive way to their real genetic situation. They were half white and half Indian and all Indian at the same time because of the way in which Americans structure racial categories. Such structuring blurs not only the genetics of intermarriage but also the differences which must exist on the biological level between one Indian population and another. Amerindians are spread thinly from the Subartic to the tip of South America and from the Atlantic to the Pacific oceans.

In the United States being "Indian" generally means having any phenotypic characteristics which are similar to the characteristics of any Asiatic population and living in a state where Indians are prevalent and mainland Asiatics quite rare. An exception to this rule may be California, where large urban Indian, Japanese, and Chinese populations keep their *ethnic* identities separate through choice of living space and custom. In New York City, on the other hand, where, in spite of a settlement of Mohawks in Brooklyn, Indians are quite rare, many Indians in western dress are likely to be mistaken for orientals. Indian identity will be assigned only if an individual happens to be one of the few "typical Indian" types with a large nose, an image burned into the minds of my generation of Americans by the buffalo nickel. In the eastern social milieu an individual who is of mixed Indian-white origin is likely to be taken for a Eurasian. In effect, the definition is cultural, not biological.

An individual is also an Indian by definition if he is a registered legal inhabitant of an Indian reservation or is on the official roles of the American Indian Service. This will hold true even if he has no phenotypic characteristics remotely suggesting Asiatic traits. An individual may be phenotypically white and behaviorally Indian. As a student at the University of Wisconsin I had occasion to meet a blue-eyed, blond Shaman with a Polish last name who believed in and played a part in the local Indian religion on a nearby reservation. In some states it is good to be an Indian. It adds to one's prestige, particularly if the amount of so-called Indian blood is minimal. That is, it is nice to have a distant relative, preferably a great-grandfather, who was a member of "such and such a tribe." "Original American" ancestry might even turn out to confer some political advantage in certain areas; but in states where the Indian population is large, it is more often a disadvantage, at least among non-Indians.

There are, then, at least three ways of being Indian in the United States: (1) the legal Indian, a reservation member, who in his lifetime may lose both his legal and racial identity by moving off the reservation; (2) the phenotypic Indian, who may or may not be so identified depending on where and under what conditions he is living; and (3) the self-identified Indian who, although he may have no characteristic phenotypic traits, claims membership within what has been defined as a race. It must be noted also that certain Americans claim Indian ancestry but paradoxically consider themselves to be whites racially. In other words, genetic ancestry is not necessarily used to define racial status!

The identification of the Negro in American society follows different rules. Just as the definition of Indian depends upon a series of historical events relating to the traditional relationship between whites and Indians so the definition of the Negro also depends on cultural-historical factors. Since there is no legal obligation on the part of the United States to care for blacks as wards of the government and because black ancestry, until recently, rarely provided political advantage, phenotypically "white" individuals had little to gain from claiming African ancestry. Although black nationalism has always had a small following in the United States, and is now an important force among militant blacks, few individuals in the past took advantage of their "African ancestry." One famous exception was Walter White, who from 1931 until his death in 1955 served as the secretary of the NAACP. Until recently, most phenotypically white individuals with African ancestry tended to conceal it and pass into the white community. They redefined themselves, in social terms, as whites.

Biologically speaking, the "black man" in America represents a hybrid population composed of genes from Africa (primarily West Africa) and Europe. Thus, the genetic structure of the black population of the United States differs considerably from those African populations from which it sprang. It must also be remembered that American blacks constitute a sample of only a few of the many populations of the African continent.

The category "Negro" in the United States includes individuals who have any of the common phenotypic traits associated in the American folk taxonomy with African ancestry, or any individual who claims to have, or can be shown to have, such ancestry. Unlike Caucasians, who under certain circumstances can remain white and be part Indian too, it is almost impossible in the United States to remain white if African ancestry is found somewhere in the family tree. A very peculiar second-rate literature has grown up around this type of racial identification. Typically, a young and beautiful southern belle, about to marry the aristocratic descendant of white slave owners, suddenly discovers that she herself is a Negro through genealogical ties to some long-forgotten distant relative. She gives up her future husband and responds nobly to the call of her race by setting off on a new career caring for poor black orphans. Her marriage to a white man would be unthinkable and so a happy ending is built upon racial memory and heroic sacrifice. (Many of these stories become Hollywood films, viz., "Pinky.") One of the persistent biological myths reflected in these stories is the commonly held belief that a white–light Negro marriage might produce dark, even black, children. While two individuals of mixed ancestry can, in fact, have children somewhat darker and somewhat lighter than either parent, a marriage between a "pure" Caucasoid (whatever that is) and a Negro cannot produce offspring darker than the darker parent. The myth of black babies coming from such marriages probably arises from the fact of marriage between phenotypically light Negroes and individuals who were wrong in assuming that they had no African genes! This could happen in the south where a good deal of nonmarital intercourse took place (most often involuntarily on the part of blacks), particularly before the Civil War. Genes from such unions, some of them undoubtedly African in origin, got into the gene pool *socially* defined as white. It is for this reason that no southern American can really be sure that he is not "Negro," at least as he himself would define the term. This produces a cultural paranoia derived from social rules and definitions, as well as misunderstandings of biology.

Let me remind the reader, however, that Africans and Europeans exchanged genes for centuries before the discovery and settlement of America. Therefore, there are no pure Caucasians or pure Africans anyway!

Race, then, is clearly a folk category. The criteria used for placing individuals in one or another racial group are determined not by biological principles but by current ethnically defined rules. In various times and under various conditions the term "race" has been applied to individuals and groups carrying (1) a certain aggregate of phenotypic traits; (2) groups whose identities rest primarily on religion and common culture (Jews); (3) nations (the Irish race); and even (4) linguistic affiliation. In addition, certain terms are borrowed and misapplied to peoples who have no possible identification with the chosen category. Thus, for example, Hitler chose (with historical precedent) to emphasize the Aryan myth of German ancestry. The only scientific meaning of Aryan is linguistic, for it refers to a subgroup of the Indo-European language family. The European representatives of this linguistic category are the gypsies, a people classified as inferior by Hitler and condemned to death along with the Jews in the gas chambers and concentration camps of the Nazi era.

In Europe, the current accepted meaning of the term "race" has a built-in confusion, since ethnic identity is frequently taken as the equivalent unit. If I wanted to know the tribal identification of an individual in French West Africa, I had to ask "De quelle race êtes-vous?" An individual would respond with the name of his ethnic group. Although "race" in French and German has ethnic identity as its primary meaning, the term also carries biological connotations. This semantic confusion tends to reinforce the belief that race and culture are biologically intertwined and determined.

The fact that phenotypic characters are linked only peripherally with racial identity is demonstrated not only by American Indians but also by Jews. Studies of isolated Jewish populations from the Middle East and various parts of Europe show a considerable degree of disparity between the gene frequencies of each group. The physical variety of Jewish population is easiest to discern in Israel. One finds individuals ranging from dark skin with kinky hair, so-called Mediterranean types, to "typical Anglo-Saxons." Linguistic confusion also occurs in Israel, for Jewish settlers from England or the United States, no matter what their phenotypic characteristics, are referred to as "Anglo-Saxons."

Because the American Jewish community is mostly of European origin and because it represents a wide range of phenotypic characters, an American may not be aware of the physical differences among groups of Jews. In France, on the other hand, three distinct Jewish communities exist with almost no intermarriage between them. The first two, established French Jews and Eastern Europeans, are distinguishable on the basis of custom rather than physical characteristics; but the third group, by far the largest, is both culturally and physically distinct, having its origin in the former French colonies of North Africa.

There has always been a good deal of intermarriage between Jews and non-Jews. This has occurred in spite of endogamous rules within and outside the Jewish community. It is not surprising, therefore, that different Jewish populations bear more similarity to local populations than to Jews as a whole. There is also convincing evidence that Jews were never, even in the distant past, a genetic entity unto themselves, but constituted part of a much wider Middle Eastern population. Nonetheless, there are many individuals who are sure that they can tell a Jew from a non-Jew on the basis of phenotypic criteria alone. Such criteria are necessarily vague but generally include traits which can be used with minimal success to type individuals from the Mediterranean region, which, of course, includes Arabs. An Eastern Mediterranean in the United States is certainly more likely to be Jewish than Arab, given the proportion of Jews to Arabs in the American population. On the other hand, since Arabs are less mixed with northern Europeans, many Arabs are likely to look more "Jewish" than most Jews!

Midwesterners in the United States often confuse Italians with Jews and vice versa. Mediterranean peoples fit the Jewish stereotype, and relative unfamiliarity with Jewish and Italian cultural differences facilitates error. Some Americans take almost any New Yorker for a Jew since they define Jews not on the basis of phenotype but by behavior, particularly speech patterns. The New York accent is often mistakenly identified as a second-generation Jewish accent.

In societies with significant socio-racial differentiation, the dominant group usually determines the defining characteristics for all subdivisions, even though a great deal of ambiguity may exist in the system. This is certainly true in the United States. Every individual who can be shown to have African ancestry is considered to be a Negro, although the legal definition of Negro in the United States differs according to various state laws. If the social situation were reversed, that is, if American Blacks constituted the majority and a similar kind of socioeconomic relationship between the two groups existed, it is likely that the category "white" would include all individuals who could be shown to have some element of European ancestry. It is even possible that many black nationalists who today advocate total separation of the races would find themselves excluded from their own utopias. The solidarity of "Negroes" in the United States, as far as it exists, is a product of social conditions which, if changed, could produce new social cleavages.

In the past an American Negro could escape from his identity only if he were acceptable as phenotypically white by the white community. Such an individual could remove himself from the Negro social milieu, move to a white neighborhood, give up certain cultural practices, and from then on be white. Ethnic identity tends to be strong, however, and some who passed into the white community felt culturally deprived. One way out of this dilemma was to operate as a part-time white in the business community but to remain Negro socially and culturally after hours. The rise of black nationalism and black pride has now produced a movement away from the imitation of Caucasoid phenotypic characteristics, especially among youth. Many individuals attempt to look as phenotypically African as possible. Those American Negroes who still choose to pass are free to decide between two options. They may, if they are phenotypically white, continue to pass into the white community in the traditional manner, but if they are black they may become "African" rather than American Negro. This phenotypic change is accomplished by the adoption of Afro hair styles and by the wearing of traditional African dress. There are many instances in which upper-middle-class Americans, particularly in the sophisticated New York area, are happy to open their homes to genuine and sometimes not so genuine Africans, while they would be horrified at the thought of entertaining American Negroes. I know cases in which wealthy Long Island matrons offered to entertain Africans for a day or even for a weekend. But they always emphasized that the individuals sent to them by international agencies should be dressed in "native" costume. After all, they don't want their neighbors to get the wrong idea, and what would the black servants think! Such is the schizophrenia of the American racial scene.

The Puerto Rican community in New York City, which ranges widely in phenotypic characters, is relatively solid in the face of prejudice which it encounters from both whites and blacks. Puerto Rican identity is based upon culture and language rather than on biological characteristics. Most Puerto Ricans, even the most black, reject identification with the Negro subgroups. When filling out employment forms which request racial type, most respond with "Puerto Rican" or "Spanish." Both the white and Negro communities in New York appear willing to accept this definition as long as an individual speaks Spanish or English with a Spanish accent. It is the accent in this case which supplies the appropriate label. Second- and third-generation Puerto Rican New Yorkers are going to have considerable difficulty preserving this identity. The lighter ones will be able to slip quietly into white society; the others may well become blacks unless they maintain active ties with the Puerto Rican community.

Other Latin Americans and Spaniards often resent identification with Puerto Ricans. But since the defining characteristic is linguistic, there is little an individual can do about such misidentification when it occurs under conditions of casual contact.

Laymen who consider the racial identity of an individual as the basis of social interaction assume not only that they are talking about specific and discrete biological entities but also that the phenotypic characteristics associated with each of these groups are directly linked to biological behavioral differences. I once observed this type of thinking while on a family camping trip. A ranger in the state park in which we had pitched our tent referred to the local population as a bunch of lazy, good-for-nothing individuals who produced children in great numbers, preferred to live on welfare rather than to work, were congenitally immoral, and displayed a tendency toward drunkenness. When pressed to describe any good features of this group he admitted that they were generally good-natured and highly musical, with a good sense of rhythm. I am sure that the American reader, conditioned as he is by the prevailing prejudices in the United States, will recognize this portrait as a stereotype of the American Negro. Not so! For that summer my family was camped on the Gaspé Peninsula in Quebec Province, Canada, and my informant was a local Scotch-Irish Canadian (a member of a local minority) describing what he was sure were the hereditary characteristics of French Canadians.

Carleton Putnam (1961), a Northerner, commits many of these opinions to the printed page, including "the good side" of Negroes: "Those qualities of heart and mind, the sense of humor, etc." Putnam also suggests that if he were lost in the African jungle he would want a Negro with his special talents (brought about by natural selection no doubt) to come to his aid. The assumption he operates under is that individual races are all fit to live, but only in the environments in which they evolved, and that racial behavioral patterns are fixed. It would be a kindness to the Negro to restore him to Africa where he thrives, and a blessing to whites, since it would remove the danger of interracial marriages which threaten the European adaptation.

At a single point in time, stereotyped racial and

ethnic definitions may show relatively high consistency. For reasons of sanity and self-respect, however, one's own ethnic group is generally excluded from such stereotypes except where nonpejorative characteristics (which add credibility to the overall typing) are affirmed. Years ago Bogardus (1925, 1933) demonstrated that Americans from a wide range of ethnic backgrounds tend to have the same feelings toward a standard list of ethnic groups. The degree of social distance which respondents maintained between themselves and members of such groups was highly predictable. Social distance was rated on a scale including such statements as: would be willing to have in my country as a citizen; would be willing to have in my neighborhood; would be willing to have a member of my family. Outside of one's own group, Anglo-Saxon Americans were consistently rated as the most desirable. Negroes were the least desirable. At the time of the study attitudes toward minorities were amazingly consistent across a broad spectrum of the United States. While these attitudes may have changed recently, and while certain groups may have moved up or down the scale, retests employing the same measures yield results which continue to display consistent sets of responses.

While stereotypes are rather stable at any given time, a historical view of racial attitudes in America shows that they are subject to wide change and even reversals. Those who claim objective validity for such concepts would do well to look at such attitudes from a temporal perspective. Such a survey has been published by Gloria Marshall, an anthropologist. William Stanton's *The Leopard's Spots* (1960) and Thomas Gossett's

Race: The History of an Idea in America (1963) also demonstrate the rather startling temporal flexibility of these concepts. Marshall (1968) reminds us that in writing about the 1840s and 1850s

Charles F. Adams, Jr. remarked that the "Irish race," being "quick of impulse, sympathetic, ignorant, and credulous . . . have as few elements in common with native New Englanders as one race of man well can have with another."

By the 1870s the Irish, representative of the so-called Celtic race, gained dominance in some service industries in Boston and nearby mill towns; by the 1880s, they wrested the political leadership of Boston from the old New England aristocracy. The political supremacy of the Brahmins having been challenged, various academicians from this aristocracy sought to prove that the increased influx of members of the Celtic and other "inferior races" undermined the chances for the survival of democratic institutions which were Teutonic in origin and transmitted "through the blood." (157–58)

Between the 1830s and 1890s, the Celts were described as ignorant, shiftless, credulous, impulsive, and mechanically inept; they were inclined towards drinking and related crimes. By the 1890s, when the Irish were the political leaders of the hub of New England and large numbers of southern Europeans were coming to the United States, the Irish had become tolerated aliens. The shift in attitude toward the Celtic race reflected the change in the political situation. The Irish were said to have "a remarkable race trait of adaptability which explained the achievement of the more intelligent and prosperous of the Boston group." Moreover, the Irish "above all races [had] the mixture of ingenuity, firmness, human sympathy, comradeship, and daring that [made them] the amalgamator of races." (160)

It is not my intention to claim that typical patterns of behavior are not associated with membership in ethnic groups but only that such patterns are cultural in origin, rather than biological, and that they are subject to modification. There are probably good grounds for the belief that differences exist in the average personality types and behaviors of members of various culturally distinct populations. Many anthropologists have attempted to give scientific validity to this idea through the study of so-called national character. While the success in objectivizing this material has been less than spectacular, there are strong hints that national character is a sound principle. Unfortunately, such studies have been marred by poor design, particularly poor sampling (often utilizing refugee populations, most of the work being done during the Second World War when anthropologists were unable to get into the field) and by a rather naive approach to data. For example, national literature written by members of an intellectual elite for that elite has been used to make generalizations about the average citizens of a country.

Overt behavior is often highly patterned and predictable in relation to group membership. There are some societies in which the individual is expected to be shy, withdrawn, and diffident, at least publicly, and contrasting societies in which individuals who fit the pattern are expected to be rather flamboyant and aggressive. Again it must be stressed that ethnographic descriptions and national character studies in particular tend to overemphasize similarities and to play down differences between individuals. This is because the ethnographic portrait is a synthesis of a culture. Bernice Rosman, a clinical psychologist, tested a large sample of Kanuri in Nigeria with a series of psychological measures. She was struck by the wide range of personality differences in Kanuri society rather than by a set personality type. Psychologically oriented anthropologists have explained such apparently contradictory results by stating that certain core attitudes and behaviors are held by the majority of the members of a particular group, but that in other areas of personality there could be a considerably wide range of behaviors. I think that even in this respect a certain caution must be maintained when deriving such concepts as core attitudes, for such markers are undoubtedly subject to temporal change.

I write this book during a year's leave in Paris. I find that many of the stereotyped impressions of the French held by Americans, particularly American tourists, are frequently more humorous than valid. The layman tends to associate certain items of behavior with elements of his own symbol system, confusing it with the symbol systems of other people. A particularly striking case is the difference in the attitude of the French and Americans toward elimination. Americans tend to be rather prudish about exposure of the body in public. (I exclude the theater in both countries from comparison, although it might be worth remarking that in the past few years movie censorship in America has relaxed to the point where anything which can be screened in France can be screened in America as well.) One is likely to see French people, usually but not always men, urinating along the sides of roadways in full view of passing traffic. Shyness extends only to turning one's back on the audience. The occasional co-ed bathroom is a source of embarrassment and amusement to Americans, and the frequent invasion of men's rooms by cleaning ladies, unannounced, has led to many a red-faced male. I myself was startled in a Viennese railroad station (the attitude tends to be pan-European) when I found an old lady apparently established permanently in the men's room in full view of the urinals. She was selling toilet paper for a few shillings.

The frequent first response of the naive American is to assume that the French and other Europeans are somehow less moral than Americans. While French intellectuals may be more free and less guilt-ridden about sex than Americans or at least Americans of my generation, the French middle class, not to mention the peasantry, is quite strait-laced and prudish. The real difference is that the French make a separation between elimination and sex. Americans, on the other hand, have a tendency to confuse these distinct aspects of behavior which are associated with the same physical equipment. When one understands this, the behavior in question takes on a new and quite different meaning.

Racists have always attempted to convince their contemporaries that physical characteristics are somehow markers for innate behavioral differences and that these physical and behavioral differences can always be rated on some kind of superiority-inferiority scale. The dominant group is not only aesthetically beautiful but is also virtuous, intelligent, brave, honest, healthy, etc. The subordinant group is identified by the opposite characteristics. Racists forget that aesthetic standards are relative and that other standards of beauty can lead to equally pressing claims for behavioral superiority. The pervasiveness of the aesthetic approach to race is so strong that within the same overall social system a debased group is likely to adopt the standards of the superordinate group and value physical features which contrast with their own average types. When this occurs, the racist seizes upon it as proof of the absolute value of his own standards. Thus, for a long time in the United States, many who were socially defined as Negroes, regardless of their skin color and other phenotypic traits, tended to value Caucasoid physical features. Such items as bleaching cream and hair straighteners sold well, and many Negro mothers hoped that their daughters would marry light-skinned men. (Others sought darker husbands who could be controlled socially by their lighter wives.) The growing awareness of blackness as a positive virtue has tended to change this attitude among many in the American Negro community. "Afro" hair styles are common, particularly among the college educated and students, and bleaching cream is considered demeaning by many blacks. Some years ago (1960) I studied a black pentecostal church in which products were marketed under a church label. Although bleaching cream was one of these, the local pastor said publicly that no one wanted "that kind of stuff anymore." The members of this church were all recent immigrants from the South and members of the lower-class community. Yet between 1950 and 1960 they had changed their cultural-aesthetic values significantly. Products associated with the white phenotype had become at least partially obsolete.

Many social scientists attempting to explain prejudice have relied on the cognitive aspect of racial differences. They assume that racism can be explained on the basis of selective perception and related value judgments. It has been suggested by some, for example, that the attitude of Japanese toward Europeans and blacks depends upon preconceived judgments rooted in the Japanese aesthetic and mythical system. It has also been suggested that differences in racial awareness and degrees of discrimination in North and South America depend simply upon differences in attitude toward race and slavery among Latin and Anglo-Saxon peoples. These two cases provide an interesting contrast. The Japanese never experienced a social system in which blacks or whites constituted a part. In North and South America an economic system predicated upon slavery affected every aspect of the relationship between Euro-American and Afro-American populations. The Japanese case is interesting, too, because the contact between whites and yellows eventually forced a kind of self-evaluation on the Japanese. As we shall see, the direction such an evaluation took depended on the relationship individuals had to Western culture. The Japanese were able to construct two different aesthetic systems out of the same basic material. Which system they chose was not merely a function of preconceived notions about race and human differences.

Hiroshi Wagatsuma (1967) has traced the history of attitudes of Japanese toward whites and blacks as well as changes in their own self-image under the influence of contact with outsiders. His review is particularly valuable since it spans the period from the eighth to the twentieth century and contains translations of material available only in Japanese.

Long before the Japanese had any experience with either Caucasoids or Negroids, they placed a high value on white skin and associated the color black with filth and evil. Japanese self-perception varied considerably from the European stereotype of the Oriental. The word "white" (shiori) was used to describe the ideal Japanese type, and while some individuals were described as "brown" the term "yellow" was never employed.

During the Nara period (710 to 793), Court ladies used cosmetics, particularly white powder which was applied to the face. Long smooth black hair was also considered of high aesthetic value while kinky or even wavy hair was taken as a sign of ugliness and animality.

The persistence of this aesthetic is documented through early Japanese history in both novels and courtly records. It continues right up through the Tokugawa period (1903) to the Meiji reforms into modern times. So concerned were upper-class Japanese with whiteness of skin that they were careful, even compulsive, in their avoidance of the sun. Ladies who ventured forth in the daytime did so swathed in hats and wrapped in cloths which prevented even partial exposure of the skin. In addition, they meticulously cared for the skin, bathing with special water, steaming the face, and massaging the body and face with bags full of rice bran and the dripping of the Japanese Nightingale. While Wagatsuma stresses the traditional features of this preoccupation with whiteness, he is careful to point out that such whiteness in upper-class Japanese served the social purpose of insuring an unambiguous mask of distinction for the rich. They were conspicuous in their whiteness and therefore also conspicuous in their cultivation of leisure. Whiteness was a particularly successful mark of upper-class status among a people who have a rather uniform tendency to tan deeply when exposed even to the minimum of sunlight.

While this attitude was consistently applied toward feminine beauty, a certain ambiguity existed concerning males. Whiteness and leisure contrasted with the life of the warrior samurai who was also an idol of the upper class.

From the beginning of the 19th century, the Kabuki actors set the standards of men's beauty. A rather feminine type of male with a slender figure, well formed face, white skin, black hair and red lips became a favorite object of feminine desire. . . . By the middle of the nineteenth century these characteristics began to be considered effeminate. A man with a more dusky skin and piquantly handsome face became the preferred type. (Wagatsuma 1967:411)

Japanese began to distinguish between a beautiful man who had white skin and an attractive man who was dark-skinned, masculine, and dependable.

Before the sealing of Japan by the Tokugawa government, there was considerable contact with Europeans of Portuguese, Spanish, and Dutch origin. The Portuguese and Spanish were referred to as Southern Barbarians and the Dutch as Red-headed people. Wagatsuma, who has examined art from this period, points out that several pictures which include both Japanese and Europeans show Spanish, Portuguese, and Japanese men with faces of equal skin color, flesh or light brown. Japanese women, however, were depicted as white. The Dutch in the pictures were given white or grey faces, an obvious indication that the Japanese artist could distinguish between different degrees of whiteness among Europeans. But women placed near the Dutch in these pictures were always given whiter faces. Thus they were able to maintain the fiction or the perception that Japanese females were even whiter than these foreigners.

The features of Europeans which most impressed

the Japanese artist were height, hair color, general hair-iness, big noses, and eyes. The Japanese were also impressed by the physical features of some African servants of the Portuguese and Dutch: "The faces of Negroes are painted in a leaden or blackish grey, and their hair is shown as extremely frizzy. The physiog-nomy of the Negroes is somewhat caricatured and in some instances closely resembles the devils and de-mons of Japanese mythology" (Wagatsuma 1967: 413).

The latter characterization is explained by the fact that the African fit closely to the preconceived notions of devils which existed in the Japanese mythical system.

At the time of Perry's voyage to Japan both Ameri-can and Japanese men were seen as white, but Ameri-cans were distinguished from the Japanese by a greater amount of hair, particularly the beard. At the same time Japanese women were depicted as whiter than both American and Japanese men. In an attempt to maintain their aesthetic superiority, the Japanese began to pick and choose those traits which they cared to emphasize in both their own people and among Eu-ropean foreigners. A trade expedition to the United States in 1860 included sixty-three warriors many of whom recorded their experiences and views of Amer-ica. One wrote: "The women's skin was white and they were charming in their gala dresses . . . but their hair was red and their eyes looked like dog's eyes which was quite disheartening. . . . Occasionally I saw women with black hair and black eyes. They must have been of some Asian race and naturally they looked more attractive and beautiful."

Nonetheless, in the Meiji period, which saw the rapid expansion of Japanese technology, the perception of feminine beauty began to change. Wagatsuma tells us that it was not until the 1920s, however, that wavy hair was an acceptable feature of feminine beauty. American motion pictures have had a strong effect in Japan—so much so that Japanese movie stars today, including some men, alter their physical features to conform with Western standards. Thus they have plastic surgery done to lift the bridge of the nose and to do away with the epicanthic fold. At the same time, some Japanese have developed a certain degree of self-hatred, which is so characteristic of peoples who are made, for one reason or another, to adopt the social and aesthetic values of foreigners. Thus, the white skin of European women has become an obsession for some Japanese. Beautiful Japanese women are still seen as white, but only until they are compared with Europeans. Some Japanese have come to associate Caucasian features with virility: "white skin, deep set eyes, wavy hair, of a color other than black, a tall, stout, hairy body, and large hands and feet seems to evoke in many Japanese an association with 'vitality,' 'superior energy,' strong sexuality or animality. . . ." (Wagatsuma 1967:422) Thus, the Caucasian symbolizes for many Japanese what the Negro symbolizes for many white Americans!

Wagatsuma notes the remarks of a Japanese hairdresser who was impressed by the difference in skin color between whites and Japanese. "After attending to several Caucasian customers in a row, when I turn to a Japanese lady, the change in color is very striking. She *is* yellow. It always comes to me as a kind of shock, this yellow color." (Wagatsuma 1967:422) Another informant said: "My daughter is very 'white' among the Japanese. Looking at her face, I often say to myself how white she is. As a mother I feel happy. But when I see her among Caucasian children in a nursery school, alas, my daughter is *yellow* indeed." (Wagatsuma 1967:423)

Conservative Japanese, on the other hand, manage to maintain an animosity toward Euro-American culture and an aesthetic which maintains the superiority of Japanese women. Caucasian skin is perceived as rough, wrinkled, spotted, and speckled; the pores are seen as large. Finally, Caucasian skin is denied its whiteness. It is more like sausage skin, transparent rather than white.

At the same time the Japanese attitude toward blacks remains relatively stable, and highly negative. This attitude is maintained, I think, primarily because the Japanese have had little personal experience with blacks. Blackness, which is an integral part of the belief system, is combined with such negative qualities as death, vice, and despair. Black is also taken as the opposite of purity and cleanliness. Such a conception becomes a vehicle for self-hatred, however, to the extent that some Japanese identify themselves with Africans to their own detriment in comparison with Caucasians.

It is evident that the Japanese attitudes toward skin color are relatively persistent and reflect a continuity of traditional beliefs. Nonetheless, self-perception and perception of others depend very much upon the interpretation of aesthetic principles; and these, in turn, are determined by social position and degree of Westernization. Those holding traditional Japanese values employ negative aesthetic standards in reference to whites, while moderns feel somewhat inferior and stress their own darkness in contrast to the whiteness of Caucasians. In addition, they attempt to minimize differences cosmetically.

Except for the Ainu, who live on Hokkaido Island, there are no significant racial minorities in Japan. There is, however, an outcast group, the Burakumin, formerly known as the Eta. Historical records show clearly that the Burakumin are descendants of former Japanese serfs and that they are both physically and genetically Japanese. Yet non-Eta Japanese are convinced that Burakumin are racially distinct and attribute their origin to the original aboriginal settlers of Japan, to Philippine Negritoes, or to a group of Korean prisoners of war who were said to have been brought to Japan in the sixteenth century. The Eta were, and in many cases still are, segregated, living in small groups on the outskirts of towns. Their relationship to Japanese society at large is further complicated by their residence with other social outcasts.

While the Japanese do not use skin color as a defining characteristic of the Eta, other doubtful physical features are attributed to them. Some of these cannot be verified through simple observation, and the ordinary Japanese is in no position to judge their validity. These stigmata include social and physical characteristics.

(1) the practice of eating meat when the Japanese proper despise it
(2) reddish tinge in eye color
(3) prominence of cheekbones
(4) non-Mongolian type of eyes
(5) dolichocephalic head
(6) shortness of stature
(7) shortness of the neck

In the folklore the Eta are described as dirty and diseased and able to walk barefoot through dirt without having it adhere to their feet (a characteristic which shows them to be animal-like). In addition, they are said to have abnormal sexual and excretory organs and to have one rib bone missing (see Marshall 1968: 154).

Some Japanese anthropologists have contributed to the myth of Eta distinctiveness. Thus, professional opinion has been mustered to support lay values even though scientific evidence runs contrary to these values. This is not an isolated case. American intellectual history is full of examples in which scientists followed rather than corrected folk images of racial types.

The Eta provide a particularly interesting case of selective perception. Western physical anthropologists find no legitimate observable criteria for separating the Eta from the other Japanese (although certain gene frequency differences actually do occur between the two groups).

Historians have emphasized differences with regard to slavery and attitudes toward race among Anglo-Saxon and Latin peoples, particularly the Spanish and Portuguese, to account for modern differences in social relationships between the races in North and South America. Many scholars have claimed that slavery was more corrupt and pervasive in North America than it was in South America. Marvin Harris, an anthropologist, contends that these explanations rest too heavily on belief systems. He finds significant differences between word and deed in Latin America and points to instances of benevolent thought as well as action in the United States, particularly in the northern states when these areas were still populated by Anglo-Saxons.

Through a careful analysis of historical records and socioenvironmental conditions in the two areas Harris shows how socioenvironmental factors provide a much more economical and reasonable explanation than those which rely on the persistence of philosophical systems.

Harris first attempts to explain why Negro Africans were preferred to native Indians as slaves by both North and South American colonists. He suggests that "the Africans had been pre-adapted by their cultural experience to cope with the demands of regular field labor . . . the Negroes over centuries of indirect contact with North Africa and Europe, probably had acquired immunities to certain common European disease organisms which were lethal to the American Indians." (Harris 1964:14)

It is well known that smallpox and measles wiped out considerable numbers of Amerindians in the early days of colonization. Harris suggests, in addition, that the only way to effectively exploit Africa was to export its people as slaves to the New World where they could work on established plantations. The development of plantation agriculture in Africa came only in the late nineteenth century and never in any degree in West Africa, the area from which most of the slaves were taken. Such an area was considered unhealthy and unlivable for whites and it was used almost exclusively for trade involving the extraction of slaves.

In his examination of the Brazilian racial pattern, Harris demonstrates quite clearly that there was no such thing as a Negro group on the one hand and a white group on the other. Instead there existed, and still exists, a continuum ranging from "black" at one end to "white" at the other. The way in which "race" is defined leads to a multifaceted system rather than the simple dichotomy found in North America. Harris suggests that the origin of this difference is in the use of *hypodescent* in the Northern Hemisphere. Hypodescent is a system under which anyone who has any Negro ancestry is defined as a Negro. South Americans did not employ this classificatory device. Latin Americans classify racial types with a great deal of care and attention to small degrees of physical difference plus, and this plus is most important, educational and economic standing.

Racial identity in Brazil is not governed by a rigid descent rule. A Brazilian child is never automatically identified with the racial type of one or both of his parents, nor must his racial type be selected from one of only two possibilities. Over a dozen racial categories may be recognized in conformity with a combination of hair color, hair texture, eye color and skin color which actually occur. These types grade into each other like the colors of the spectrum and no one category stands significantly isolated from all the rest.
  One of the most striking consequences of the Brazilian system of racial identification is that parents and children and even brothers and sisters are frequently accepted as representative of quite opposite racial types. (Harris 1964:57)

In addition, Harris points out that the Brazilian classification system is ambiguous. Individuals find it difficult to define racial terms. Thus 40 percent of a representative sample ranked *moreno claro* as a lighter type than *mulato claro,* while 60 percent reversed this order. Individuals may also change their evaluation of someone else's racial classification at different times. In order to test this ambiguity Harris showed a sample of illustrations which varied in hair shade, hair texture, nasal and lip width, and skin tone to a group of 100 Brazilians. The results were quite astonishing. "Forty different racial types were now elicited. The highest percentage of agreement reached for any of the drawings was 70% *branco*. The lowest percentage of agreement was 18% for *sarara*. Nineteen different terms were elicited by drawing number one [sarara] and nine for number nine [branco]." (Harris 1964:58) Not only will individuals change their evaluation of the racial classification of another but a person in his own lifetime can alter his own classification through changes in behavior and economic status. The ability to pass from one sort of racial classification to another does not depend upon phenotypic characteristics nor does it have to proceed in secret. In addition it does not involve a denial of one's genetic ancestry. The color of an individual may be perceived differently depending upon his standing in the community. Harris warns us, however, that this does not mean that racial prejudice is absent in Brazil. This is a false conclusion drawn by those who indulge in a rather simple-minded comparison between Brazil and the United States. There is in fact, he tells us, a stereotype of the Negro which is quite unflattering. They are considered to be "innately inferior in intelligence, honesty and dependability." Furthermore Negroid physical features are taken to be aesthetically inferior to Caucasoid features, even by individuals who define themselves as Negroes. The interesting thing about these stereotypes is that they do not show themselves in actual behavior.

What people say they will or will not do with respect to *pretos* and *mulatos* does not issue into actual behavior. Indeed, extremely prejudiced Brazilians have been observed to behave with marked deference towards representatives of the very types whom they allege to be most inferior. Racial prejudice in Brazil, in other words, is not accompanied by systematic racial segregation and discrimination. The reason for this paradox should be clear: despite the "ideal" stereotypes, there is no "actual" status role for the Negro as Negro, for the white as white, or for the mulato as mulato. There are no racial groups. Before two individuals can decide how they ought to behave towards each other they must know more than merely that one is dark skinned and the other light. A Brazilian is never merely a "white man" or a "colored man"; he is a rich, well educated white, or a poor, uneducated white man, . . . etc. . . . The outcome of this qualification of race by education and economics determines one's class identity. It is one's class and not one's race which determines the adoption of subordinate and superordinate attitudes between specific individuals in face to face relations. . . . Color is one of the criteria of class identity; but it is not the only criterion. . . . (Harris 1964:61)

Harris disagrees strongly with the suggestion that these differences have anything to do with the humanity or inhumanity of the slaveholders. He points out, for example, that interbreeding in Portuguese Africa is extremely low in spite of a long history of colonial domination. This is contrasted with a much higher rate in South Africa among the "racist" Dutch. Before the recent developments in South Africa the white population managed to create a large mixed group known as the Cape Coloureds.

Examination of slaveholding records shows, on the other hand, that in Brazil as well as in other parts of Latin America and, of course, in North America, "law and reality had equally small resemblances to each other." The Anglo-Saxon records in relation to slaves are more ambiguous than defenders of Latin America would care to admit, since New Englanders and peoples in the North of the United States abolished slavery long before it was abolished in parts of Latin America. For Harris, "Understanding of the differences in the status free 'Non-whites' in the plantation world can only emerge when one forthrightly inquires why a system which blurred the distinction between Negro and whites was materially advantageous to one set of planters, while it was the opposite to the other." (Harris 1964: 81)

Harris argues that when the New World was settled by the Portuguese and Spanish there was a severe domestic manpower shortage at home. This made it extremely difficult for them to rapidly populate their colonies. The New World was then flooded with African slaves. On the other hand, immigration to the English colonies was massive and slaves came rather late in history. Harris cites population statistics from Virginia which demonstrate that it was not until the second quarter of the eighteenth century that Negroes exceeded 25 percent of the population. In Brazil the ratio of whites to blacks was the opposite. In 1819 there were almost as many mestizos, free and slave, as whites. By 1870 there were more "mixed" individuals than whites. This situation reversed itself toward the end of the nineteenth century. Thus:

Given the chronic labor shortage in sixteenth-century Portugal and the small number of people who migrated to Brazil, the white slave owners had no choice but to create a class of free half-castes; . . . They were compelled to create an intermediate free group of half-castes, to stand between them and the slaves because there were certain essential military and economic functions for which slave labor was useless and for which no whites were available. (Harris 1964:86–87)

This was particularly true in relation to military and police duties. The half-castes were often used to fight Indians. They were also important in a growing cattle industry. "Open range mounted cowboys for obvious reasons, cannot be slaves; nor would any self-respecting Portuguese immigrant waste his time rounding up dog-gies in the middle of a parched wilderness." (p. 87)

Harris suggests also that the bulk of the food growers who supplied the necessary calories for the one-crop plantations may have been "aged and infirm manu-mitted slaves, and favorite Negro concubines who with their mulatto offspring had been set up with a bit of marginal land." (pp. 88–89) These economic activities which could only be supplied in South America by half-castes and mulattoes were performed quite ade-quately in North America by free whites, many of whom had originally come to the New World as inden-tured servants but who, after their terms of service had expired, had been able to buy land and set up farming.

Harris's analysis appears more powerful than one which depends on different moral value systems im-bedded in the culture of North and South Americans. One might add that hostility between poor whites and free Negroes in the United States was the result of di-rect competition on the lower levels of economic ac-tivity, and that this hostility was encouraged by the white ruling class as an effective control on both groups. Moral systems bend in response to new eco-nomic demands.

A series of persistent myths concerning ethnic and so-called racial groups in the United States continue to muddy intergroup relations. Among these is the myth of superiority. The supposed *superior* intellectual capacities of Jews led to their partial exclusion from educational institutions in the United States, many of which are rated among the best and also the most liberal schools. The quota system which limited the number of Jews accepted by colleges and medical schools has all but disappeared, but was extremely common until just a few years ago. Many schools which profess an open policy in student recruitment, however, continue to exclude Jews, and other minority groups as well, through the imposition of regional quotas. They are manifestly designed to provide the school with a geographically divergent student body, but since minority group members are limited in large part to certain states, regional quotas tend to cut down on minority group admissions. Quotas against Jews in the ranks of certain professions were established out of fear that "superior" (sometimes the pejorative "crafty" was substituted for intelligent) Jews would crowd out non-Jews. The myth of superintelligence in the Jewish community is reinforced in part by some Jews who believe that they as a group are endowed with a genetic propensity for high intelligence. There is no doubt that Jews as a social entity have demonstrated rapid upward social mobility in American society.

The Jew was, to use Marvin Harris's terms, culturally "preadapted" for social conditions in America. The attitude among members of the Jewish community toward learning, which in the religious context of Eastern Europe was limited to the study of holy texts, was reoriented in secular America toward professional edu-

cation. Also, Jews coming to the United States had the
way cleared for them to some extent through a network
of assistance agencies which had always been a feature
of the Jewish community. Although Eastern European
Jews experienced some prejudice at the hands of their
co-religionists who had preceded them, many of the
former extended a helping hand to the new immigrants.
The goal of the Jewish community center in the early
days was to make good Americans out of immigrants.
These agencies, which also aided non-Jews in the slums
of New York, were responsible for the rapid Americani-
zation of a large proportion of the Jewish community.
This, coupled with the folk belief held by many immigrants to the United States, that anything was possible
in America, led to a successful era of striving and accomplishment in the American Jewish community. This
does not mean that the Jew has been successful in all
areas of American life. Jewish money and power tend
to be concentrated in the areas of banking, where Jews
are a sizable *minority,* and in the mass communications
and entertainment industries. Jews have been highly
successful in the professions, particularly medicine and
law, and in the academic community. But it has been
very difficult for Jews to enter into other aspects of
American business. Big business is overwhelmingly the
territory of white protestants.

The Negro community, often contrasted unfavorably
with the Jewish community, had no such financial or
educational base on which to build. The aspirations of
Negroes were limited by the realities of the white world
rather than their own. Most professions were closed to
Negroes. While the Jew was seen by the non-Jew as a
striver and alternately despised and admired for this
behavior (it was, after all, in one sense the American

ideal), the Negro was always defined as lazy, inept, and unqualified except for the most menial positions. The American Negro was never treated to the begrudging admiration and ambiguity which at times helped Jews to succeed. The Negro was stripped of economic resources as soon as he was taken from Africa as a slave. These resources have never been restored. There was almost no base on which anything solid could be built.

In the southern United States the existence of two separate communities provided the opportunity for some Negroes to become schoolteachers, ministers and, in some cases, even doctors and lawyers. Under these conditions Negro children were able to identify, at least to a small degree, with members of their community in the professions. This was blunted, however, by the constant reminder that Negroes occupied an inferior social position. Even those blacks who entered the professions were forced to defer to less well educated or less well bred whites. In the North, on the other hand, with its hidden racism and a single social system, the professional is an extreme rarity in the Negro community. The most common models for success are those who have made the grade in sports, in music, or in crime.

Thus there are important social-psychological differences between the Jewish and American Negro communities in the United States. Jewish children have always had models of success for their own aspirations. Jewish parents could point with pride to members of the Jewish community, often within the family, who had raised their own socioeconomic standing.

The contrast between Jewish and Negro groups argues for the effect of environment and social background on the performance of ethnic groups. In light of these social explanations for intergroup differences, racists have a tendency to retreat to mythical descriptions of the Negro in Africa. The "inherent" inferiority of the Negro is now frequently proved by the fact that Africa did not develop a modern technology or Western forms of government. While modern technological society is of course a product of Western European culture, civilization did not develop in Europe. Every one of the innovations which contributed to the rise of urban civilization took place in the Middle East among peoples who fit Carleton Coon's description of "clinal." They were, in fact, the product of genetic intermixture between Europe, Africa, and Asia. But such genetic constitution is irrelevant anyway as an explanation for the rise of civilization. It is most likely that these advanced societies developed where they did in response to environmental factors. In these regions populations were held together in river valleys by surrounding deserts. The development of stable agriculture in these rich alluvial valleys allowed for high population density. An uneven distribution of natural resources stimulated trade between population centers while agricultural surpluses freed some to toil as traders and craft specialists. It is likely that the regulation of trade stimulated the development of tally systems and eventually writing and money.

Once such a cultural complex had evolved it could then spread to new environmental zones where it could flourish but in which it was unlikely to have originated. The Mediterranean world is a marvelous area for the exchange of ideas as well as genes. It is a relatively calm lake with good harbors dotted around its circumference. Boat trade in the circum-Mediterranean area began early in human history. The exchange of ideas must have been one of the crucial factors in the societal development for it allowed men to take full advantage of their creative capabilities.

Africa south of the Sahara was cut off from much of this influence. Geographically it was a poor area for contact. In the north the desert could be crossed and was crossed by traders, but they controlled what would get in and what would get out of coastal Africa. For centuries the trade consisted mostly of slaves, salt, and gold. The African coast is almost devoid of good harbors and the open Atlantic cannot be compared with the Mediterranean in terms of hospitality. The rivers are navigable only for short distances from the ocean. In spite of this, civilizations developed in west Africa in response to the trans-Saharan trade. These empires developed around cities built at strategic intersection points between traders coming from the north and those coming from the south. From the sixth century A.D. to about 1500 these empires waxed and waned as the struggle for control of trade routes continued. In many ways the culture of these empires was on a par with the Europe of that time, and many of their social institutions were similar in structure to their European counterparts. The voyages of European explorers in the fifteenth century around the coast of Africa which led to the establishment of direct trade short-circuited the older empires and most of them withered away. Imagine what would have happened to the culture and civilization of Venice if it had suddenly been deprived of its trade. Coastal trading between Africans and Europeans stimulated the development of new states in Ghana, Nigeria, and the Congo.

The educability of Africans need not be demonstrated. Opportunity and proper incentives have already transformed so-called primitives into men and women fully in tune with the technological age. Lives have changed overnight from one type of existence to another under the stimulation of acculturation. This is not a question of genetic adaptation, but simply one of learning.

The assumption on the part of European colonialists that most Africans were basically inferior left a peculiar stamp on recent history. The French in their sphere of influence decided that people from Dahomey were more intelligent than the general run of Africans. The English appear to have made the same decision in relation to the Ibo peoples of Nigeria. Indeed, factors imbedded in the social structures of these groups undoubtedly had some effect on their response to opportunity, but the decisions of the French and English are largely responsible for the success of these ethnic groups in colonial administration.

In both cases the favored position which individuals of these ethnic groups enjoyed led to a considerable degree of animosity on the part of less fortunate Africans and eventually to discrimination against them. In 1958 race riots directed against Dahomeyans broke out in the Ivory Coast. Most of them were eventually expelled from the country. The tragic war between the federal government of Nigeria and the Ibo state of Biafra had its roots, in part, in the resentment felt by non-Ibo Nigerians toward the Ibo peoples.

It is quite common in the United States to grant genetic ability to the Negro for music and dance or, to be more precise, for some types of music and some types of dance. Blacks are said to excel at rhythm. After all, is not jazz the product of American Negro culture?

While it is certainly true that jazz is African music re-born under the influence of American black (and white culture), musical sophistication varies tremendously from individual to individual, and, more importantly for this argument, from society to society. The cradle of music in Africa is clearly the West. In fact, West Africa, particularly the Guinea coast and the Congo, is the region of high artistic production in the plastic arts and dance as well. But musical ability is distributed among individuals in these populations in the per-centages one might expect elsewhere. Everyone loves music, or nearly everyone, but only some are good performers. Thus, to take an example from my own field work experience in the Ivory Coast, in a village of slightly over 100 individuals there were three good drummers—and I was one of them. Genetics may play a part in all types of musical ability, but training and emphasis on music in the culture and/or the family are also important. My mother is a professional dancer with an impeccable sense of rhythm. Is my rhythmic sense inherited? Perhaps, but I will never know, for my mother spent many hours when I was a young child training me in eurythmics. At the age of five I was the best pot and pan drummer in the neighborhood.

I made another discovery in Africa which would sur-prise many Americans. While music is loved and appre-ciated by almost everyone, these skills are admired ex-clusively in the young, except in the case of religious specialists who have to dance or drum as part of their ritual obligations. Older people who frequently indulge in musical performance are looked upon as a bit frivo-lous. The quality most admired in mature individuals, particularly but not exclusively men, is the ability to speak well. Oratorical contests are a major element in both entertainment and education among the people with whom I spent several months in field work. My tape recorder created a sensation as people clamored to hear themselves speak, attempting at each recording session to perfect their oratorical abilities. I might add that the qualities demanded of a good orator include the ability to speak without mistake or hesitation, and a high standard of perfection and originality, particularly in the use of metaphor. Many West African languages are wide open to rich metaphorizing, and speech-making becomes a kind of living poetry. It seems to me that this side of African culture is reflected to some extent in the freewheeling, extemporaneous sermons so characteristic of the revivalist church of America. As such it may be a *cultural* survival.

I am not suggesting here that one should substitute the hypothesis of genetic superiority in oratory for a supposed musical gift but only that many Americans are ignorant of a culturally significant aspect of African aesthetic life. The fact that so many black children in the United States turn out to be good dancers and/or good musicians is no doubt based upon a culturally derived love of music. It is also undoubtedly based upon expectations within the white community and upon the fact that, while music was encouraged by white plantation society, oratory was discouraged. Simply put, it is easier for black Americans to achieve success in music than in forensics.

The fact that African culture has produced exceptional works of plastic art has been recognized for some time by the intellectual elite of the West. African art was, in fact, one stimulus which contributed to the shaping of the modern art movement, particularly at the turn of the century among such painters as Picasso,

Modigliani, and Braque. This fact, however, has been greeted with mixed emotions by some Westerners. Arnold Toynbee, for example, once suggested that the impact of African art on the West was a sign of fatigue and degeneration in Western culture. This, of course, is pure nonsense. Artists have always been stimulated by the work of others both within and beyond their own aesthetic systems. When old forms and techniques are played out, new stimuli produce a fluorescence which involves experimentation with alien forms and their eventual reinterpretation into particular traditions.

Some individuals, particularly prewar German anthropologists, refused to accept the fact that Africans could produce realistic art. The bulk of African sculpture is either abstract or expressionistic (to borrow Western terms), but a good deal of first-rate realistic art also exists. The most beautiful examples come from Ife and early Benin sculptures in Nigeria and Dahomey. When faced with such examples, these German anthropologists attempted to rationalize their way out of a self-created racial dilemma by calling forth Greek influence on early West African culture!

A look at the field of dance shows that blacks have performed well in certain areas and rather poorly in others. Again, the correlations are directly linked to existing opportunities controlled by the white community. Thus, Negroes were accepted very early as jazz dancers and also in modern dance, which came into being in a climate of left-wing egalitarianism. In this field a series of integrated dance companies developed and continue to perform. The training in modern dance is at least as rigorous as that in ballet, and much the same physical and intellectual skill is required. Nonetheless, few blacks have been successful in ballet and those who have made it have done so recently, in most cases with governmentally subsidized companies. One can find the same parallels in singing. Few Negroes have been successful in opera although some have managed to reach the concert stage. In general, ballet and opera represent the establishment. More often than not the boards of ballet and opera companies are, at the least, conservative. It has been extremely difficult for the Negro to enter this side of the performing world. Clearly, the failure has to do with institutions and not with differences in the *type* of musical or dance talent "inherited" by individuals of African ancestry.

The same can be said for sports. Where institutions have been willing to admit Negroes, they have excelled. Track and field have been open for some time, basketball and baseball for a much shorter period, and tennis the least. Once a sport is open it is invaded with an energy suggestive of "genetic" talent. Remember, however, the case of the Jews. Professionally restricted minority groups flood into open areas. The distribution of blacks in sports as well as other areas of American life will undoubtedly even out with the elimination of prejudice and the development of true equal opportunity. Certainly the small *average* physical differences between the length of the lower limb (with the average Negro having a longer limb than the average white) is not sufficient to account for success in track and field. The outstanding performer in sports is not the average white or Negro in any case. And if one is carried away by the accomplishments of black athletes, remember that American whites and blacks alike tend to outperform African athletes when they face each other in competition.

Such common distortions of reality lead us to question the methodology employed in sorting out differences in behavior among individuals and social groups. There are two general principles of universal scientific application which I think have been largely overlooked by scientists and laymen alike when approaching the question of race. These are (1) the null hypothesis and (2) Occam's razor.

The Null Hypothesis

The null hypothesis—the basic starting point for all scientific investigation—states that there is *no* relationship between sets of phenomena or sets of variables. That is, given A and B, there is no relationship between them. The experimental task, then, consists of disproving the null hypothesis, that is, demonstrating real relationships between variables. The null hypothesis is based on the philosophical and physical principle that there is a tendency toward disorder in the universe. This is also referred to as the law of entropy. It is also based on the fact that it is impossible to prove that some kind of relationship does not exist between variables. This has been used advantageously by racists who often remind us that no intellectually honest anthropologist or psychologist has ever stated that the races are equal in intellectual capacity. While some racists might admit that tests of intellect cannot separate genetic and environmental effects, the door is left open. "The scientific community admits that there are doubts about the equality of race." Remember the null hypothesis is *accepted* until proved false; it is never *proven* true. The race-intelligence issue stated in terms of the null hypothesis is: There is no relationship between phenomenon A (race) and phenomenon B (intelligence). Those who believe otherwise must disprove the null hypothesis, something that has never been done satisfactorily. It is unfortunate that so many scientists have responded coyly to questions about race and intelligence and it is equally unfortunate that they have been led to a defensive position; for with the null hypothesis behind them, they can turn the question around and demand that racists offer *convincing* proof that the hypothetical relationship does in fact exist.

It is of no consequence whatsoever that anthropologists, including this one, refuse to state categorically that membership in an ethnic group or population is unrelated to intelligence. Such a statement would be in violation of our own methods, but it in no way implies that the profession as a whole suffers from lingering doubts. From where could these doubts arise? We are dealing with two constructs which themselves appear to be less and less of a biological fact. Intelligence is at the very least a complicated concept hidden behind simple definitions.

Occam's Razor

Occam's razor simply stated is: "giving explanations we must not suppose that more things exist than we have evidence for." If two explanations for the same set of phenomena exist and if both meet the requirements of logical and empirical proof, the simpler of the two shall be accepted. Let us apply this principle to problems of intergroup differences. Occam's razor can be applied to questions concerning the priority of sociological or biological explanations of differential group behavior when both are offered. The sociological explanation is typically simpler than the usually convoluted, even baroque, genetic explanation of differences between people.

Let us look, for example, at the intellectual history of different national and ethnic groups to see whether sociological or biological explanations of performance are the more adequate and simple. A glance at Europe shows us that different peoples made significant cultural contributions at different times, sometimes centuries apart, and that the particular intellectual discipline varied with time and region as well. Greek philosophy and science flourished around the fourth century B.C. and then faded rather rapidly. Are we to assume that the infusion of barbarian genes into the Greek population snuffed out their creative instinct? Such explanations have, in fact, been offered, but they suffer from the fact that the same barbarians, Northern Europeans, mixed perhaps with some Asiatic populations made their own contributions to art and literature at another date. Remember also that much of Renaissance learning and art developed not only from the discovery of Greek and Roman forms but as the result of stimulation from the Arab world. Looking past the Renaissance on the European continent we find a remarkable flourishing of literature and music in Elizabethan England. Literature, which began as a major art before the Elizabethan age, continued through the centuries; musical development faded rapidly. After 1600 music became largely a German, Italian, and, to a lesser extent, a French art. Italian painting developed rapidly in the fourteenth, fifteenth, and sixteenth centuries but soon turned toward an overly complex rococo style. This did not occur before the Italians had made their mark on Flemish painters who, stimulated by the Southern Renaissance, brought Italian style into conformity with their own cultural values. Are we to explain these developments, the flowering and sudden declines, on the basis of changes in gene frequency? Must we hypothesize an inflow of new genes every time a people begin to produce great art and another inflow when cultural output declines? The more reasonable explanation would be that the flow of ideas stimulates new types of local creativity which are experimented with, developed, and eventually wrung out. Such an explanation not only is simpler but fits the data better. Much historical evidence exists allowing us to trace the path of artistic accomplishment in Europe. This path was widened in the early twentieth century with the discovery by Europeans of Asiatic and African art. The influx of new ideas and material from all over the world served as the basis for a new thrust of creativity, a new game of reinterpretation, the outcome of which was a new national and international art style. The most recent product of this process of borrowing and reinterpretation is manifest in modern popular music. The roots of Rock music lie in Africa, but its branches developed out of black culture in America. Recently, strains of Middle Eastern and Indian music have been added. The so-called museum without walls of Malraux exists not only for plastic art and painting but for music. Just as the art book has widened our horizons, the technology of the recording industry has exposed us to a range of exotic music from all parts of the world. The exchange is cultural, not genetic.

Turning to politics, must we postulate that the development of stable democracy in Switzerland is the outcome of a perfect blend of genetic material from Germany, Italy, and France, three countries which have had considerable difficulty preserving individual freedom? Was it the small genetic contribution of local Alpine genes which tipped the balance in favor of stability? Why are French Swiss so different culturally from Frenchmen across the border? These are all historical questions with reasonable historical answers. Yet I do not ask them to set up a straw man. A biological answer was offered in the nineteenth century by the French litterateur, Gobineau. His ideas deviated somewhat from the pure race theories. Perfection in national character, he thought, could only come about through the perfect blending of genes from different populations. Each race had its contribution from one or another group. Gobineau's theory supported the declared superiority of the French nobility, which was assumed to differ genetically from the masses. It was essentially an attempt to support aristocratic government against democracy, but it was greeted with favor in Germany and distorted into the mold of pure-race theorists. Thus, the Germans, who were not as Teutonic as Gobineau would have liked, were transformed by such racists as Wagner, and later Hitler, into the Master race.

The assumption that national character is immutable and tied to the genetic history of a population leads to strange political theories indeed. For if this were true we would have to accept the fact that mankind is the prisoner of his biological destiny and that the destiny of each group is different but predetermined by genetic structure. If this were the case, then what would anthropology and history have to teach us?

Prejudice is easy. Tensions develop in the context of all social relationships and they are most easily dissipated through aggression aimed toward some "out" group. Often it is difficult to find objective differences among peoples. No matter, they can be created. A social group can be transformed into a biological entity by ascribing physical anomalies to it. The Eta of Japan have their peculiar sexual organs, and for some Europeans so do all Asiatics! The Nazis were sure they could tell a Jew from a non-Jew and yet Jews were forced to wear yellow stars. You can almost always tell a Negro from a white in the United States, but in Latin America the problem is more complicated. It is a case of different social definitions and not different genes. Witches were said to have had physical features which distinguished them from the ordinary mortal. Spots on their bodies were taken as supernumerary mammary glands, but who do not have spots on their bodies? When one human group needs to stigmatize another, it is not difficult to find some feature or some imagined difference to create social distance.

And what is more difficult and more disheartening is that people may not be aware that they are prejudiced. "Some of my best friends are" is the well-known prelude to a self-admission of prejudice phrased as a denial. I began this chapter with a personal experience and I should like to end it with one which will serve to illustrate this last point.

During the course of my field work with the Abron, a people of the eastern Ivory Coast, I was frequently asked about black-white relations in the United States. The Abron were particularly impressed by the fact that more black people lived in the United States than in their own country. Many of the youths were familiar with stories of race prejudice in America, for they had picked them up in their history classes. One day as I sat in the village square talking to a group of young men home on vacation from the Lycée Technique in the capital city of Abidjan, the subject of American race relations was raised again. "Is it true," they wanted to know, "that segregation is a fact of life in many parts of the United States." I attempted as best I could to explain the situation to them, and presented what I thought was an honest picture, including the fact that there were many Americans who abhorred the situation. But they could not comprehend such a system. It was just as if I had been explaining Martians to a group of New Yorkers. I suddenly realized they had a similar phenomenon in their own culture. The Abron are an agricultural people, who (in addition to subsistence crops) raise coffee and cocoa for the national and international market. They are quite well off for a people in an underdeveloped country. They also suffer from a labor shortage. They must hire members of other ethnic groups as plantation workers. Most of the hired hands among the Abron are Mossi, an ethnic group from the neighboring country, Upper Volta. The Mossi live in a rather barren environment, and although they have been organized as a state for centuries (they are a vestige of one of the great sub-Saharan kingdoms) they have a surplus population which must move off the land and seek work elsewhere. The Mossi migrate in great numbers to Ghana and the Ivory Coast where they find work as agricultural laborers. The Abron know them as poor, somewhat rootless people. They have no knowledge of the Mossi Empire or Mossi history and they treat the Mossi with a good deal of contempt. Mossi are segregated in certain parts of the larger Abron villages although they do live among Abron families in smaller towns. They dig and use separate latrines and generally keep to themselves. Few

Abron ever bother to learn the names of Mossi in their village unless they happen to be working as one's own hired hands. If they want to talk to a village Mossi, they say "Hey Mossi." One day while I was treating a Mossi youth for a bad infection I noticed that he did not flinch when I poured alcohol on the wound. I turned to my Abron assistant and said, "This man is very brave." "Oh no," my assistant said confidently, "the Mossi don't feel pain." On another occasion I spent an afternoon recording Mossi songs. A large group of admiring Abron gathered around. After the session one of my friends said, "These Mossi are really talented for music." In our conversation about racism in America I brought up the Mossi in an attempt to explain that prejudice was not limited to my own people. But one of the young men protested, "We cannot be prejudiced because we are all black." Another agreed, but added, "And besides, some of my best friends are Mossi."

❀

Chapter 7

INTELLIGENCE,
INTELLIGENCE TESTS,
AND INTELLIGENCE TESTERS

IN THIS final chapter I shall discuss comparative group intelligence, a subject which serves to close the argument of this book. It is here that confusions between sociological and biological definitions of race are most destructive and misleading.

To begin, it must be made clear that we are dealing with two nebulous concepts, race and intelligence.

What is intelligence? If one chooses to follow the hard line of American behaviorist psychology, the definitional problem can be sidestepped, at least temporarily, by defining intelligence operationally—as what intelligence tests measure. Tests designed to measure intelligence are, of course, structured in relation to some theory, but I wish to delay discussion of this problem for the moment. Let us begin with the test results and work backward to their origin and the concept or concepts of intelligence they reflect. Suffice it to say here that intelligence is a comparative phenomenon; test scores can be ranked. When rankings are ordered and compared in terms of group membership it appears that certain sociologically defined units (some of them misnamed races) produce scores which on the average are consistently below or above that of the standard white American range. These substandard groups include American Indians, Negroes, and various other ethnic groups, particularly those with a history of recent immigration to the United States. Class breakdown of scores shows that middle- and upper-class whites score better than lower-class whites and that people in the north of the United States score better than those in the south. Much has been made of the fact that northern Negroes scored better on one type of test, the Army alpha given around the time of the First World War, than southern whites. Such results have been used to indicate parity between two separate biological groups. The lower scores for Negroes in the north than for northern whites are usually attributed to differences in environment and opportunity. The established differences are used to bolster arguments about what are biologically indefinite groups. The results really should be rephrased in terms of four *sociologically real* groups responding to test protocols given under certain specific conditions. These four groups are: southern sociological "whites"; southern sociological "Negroes"; northern sociological "whites"; and northern sociological "Negroes." Since the constitution of these groups is not controlled genetically, they have no reality as biological units and cannot rationally be treated as such. This, of course, leaves the problem of Negro intelligence or Indian intelligence in abeyance, but in my opinion this is a false question anyway and I beg the reader to be patient as we thread our way through the morass of confusion which surrounds this topic.

Let us return to the concept of intelligence itself for

a moment. What is it? Binet, one of the pioneers in this area, suggested that "intelligence was the ability to select and maintain a definite direction, the ability to make adaptations leading to a desired end and the ability to criticize one's own behavior." Spearman reduced it to the ability to deduce relations and correlations. Thorndike regarded it as the power of making good responses from the standpoint of truth and fact. Terman defined it as the ability to abstract. These definitions imply that intelligence can be measured on the basis of the rapidity of accommodation or adaptation through learning and conceptualization to unique environmental situations. The complexity of these situations and the speed by which accommodation takes place are both factors in intelligence.

Immediate difficulties arise. The selection and maintenance of a definite direction and the ability to criticize one's own behavior contain elements which are subject to modification through the socialization process and by factors of individual psychology which might tend to increase or mask inherent capabilities. A hesitant, demurring individual might do less well on tests than a more confident one even though both might have the same potential abilities. And, of course, when we deal with test results the testing situation must be considered for it is bound to be influenced by the individual's cultural and psychological background. How important this is will be seen later. Although intelligence tests are supposed to be self-contained units, i.e., units which contain all the information necessary to make judgments within the context of the test, the intellectual background, interests, and experience of the individual tested appear to have significant effects on the response. In addition certain cultures have a tendency to treat truth or fact in very different ways. In our own society Aristotelian logic is imposed early on our children, while in the East not only is a paradoxical type of logic taught but intellectuals strive to change their basic thought patterns so that they can come to accept a kind of inversion of what we would call truth. The obvious is always false; truth often lies in opposites.

Tests themselves are subject to all kinds of artifactual errors, that is, interfering factors which produce consistent differences in response but which make their interpretations either inaccurate or completely wrong. Among these are the cultural factors of attitudes toward testing in general, the amount of test sophistication an individual brings to a particular experiment, and the structure of the tests themselves which may or may not measure what the experimenter assumes they do. The tests are poor as measures of biologically based group difference because the independent variable is social, not biological, and because dependent variables include such items as motivation, intellectual background, and other psychological factors which are not controlled for in the experimental situation.

Let me summarize those factors which I think go into the intelligent (successful) response to environmental stimuli. First of all, the nature of the stimulus or cue must be considered. While there is good evidence that the ability to respond to specific types of cue is partially inherent (humans in general have good vision and hearing, and relatively poor olfactory senses), there is also evidence that a good deal of learning goes into the process as well. Perceptions are selective (they must be in order to reduce the noise to which all organisms are exposed). The response to a cue involves such psychological factors as perceptual acuity, ability to discriminate (which is most likely partially hereditary and partially learned), and ability to generalize, i.e., to form classes of data from a range of sense perceptions. This, too, is clearly a process which involves both learning and heredity. Accurate responses involve memory and the ability to retrieve necessary bits of information which are employed in problem solving. Interest and span of attention, which are highly dependent upon social and psychological factors, speed of response, and effectiveness of feedback from behavior so that corrections can be made in strategy are all important variables. No single gene, of course, could underlie all of these (and other) psychological processes. In addition each variable, dependent as it may be on a hereditary base, must be subject to environmental modification in different ways. Divergent behavioral phenotypes could emerge from the same basic genotype through environmental shaping just as similar phenotypes could arise from different genotypes conditioned in different ways. Variation in any one of the listed variables creates its own set of problems in relation to ontological development. Individuals as separate products of genetic and environmental interaction can undoubtedly improve their test responses and, what is more important, the ability to deal with real-life problems, through processes specific to their own biological and social backgrounds.

Let us return to those "bright" and "dull" rat strains developed by Tryon. Remember that later experimentation demonstrated that the rats were reacting to specific tests and that environmental factors could also have a strong effect on the performance of these inbred strains. More specifically, in three out of five maze measures dulls were either equal in performance to or better than the brights. According to the analysis of specific aspects of behavior, an analysis which goes beyond simple maze performance, brights were seen as more food-driven, low in motivation to escape water, timid in open field situations, more purposive, and less destructive. Dulls were not food-driven, they were better or average in motivation to escape water, and they were fearful of mechanical apparatus features. Note how important these facts are to a full understanding of Tryon's experiments!

Intelligence tests are designed to measure a series of abilities: for example, spatial relations, reasoning, verbal fluency, and numerical ability. But these are no more culture-free than the concepts behind them. For although we can define intelligence operationally as the ability to achieve high scores on intelligence tests, we must never forget that certain socially significant concepts lie behind the operational definition. The major concept relates I.Q. to academic performance under existing forms of education. But it must be remembered that our system is geared to middle-class success, not necessarily innate ability. Jensen (1969:9), in a review of the literature on I.Q. testing, points out that many of the psychological properties which contribute to response potential intercorrelate, even though specific tasks such as spatial relationships, verbal analogies, and numerical problem solving might bear no superficial resemblance to one another. He notes further that Spearman separated out a factor g which accounted for "general intelligence." Such an analysis led Spearman to define intelligence as the ability to deduce relations and correlates. Nonetheless, Jensen, himself unsatisfied with a unidimensional concept of intelligence, suggests two genotypically distinct basic processes which he calls level I (associative ability) and level II (conceptual ability). Level I is related to the formation of associations between related stimuli; level II, to concept learning and problem solving. Level I learning is closely associated with I.Q. in middle-class but not in lower-class children. The latter often perform well in associative tasks, but score low on I.Q. tests. According to Jensen, level I and level II abilities are inherited. Observed distributions of test scores are explained on the basis of differentials in the population and are taken as a function of social class. (Jensen 1969:114) This has occurred, according to Jensen, because of a selective educational process in which level II is the more important factor for scholastic performance.

Cohen (1969) identifies two conceptual styles, relational and analytic.

The analytic cognitive style is characterized by a formal or analytic mode of abstracting salient information from a stimulus or situation and by a stimulus-centered orientation to reality and it is parts-specific (i.e., parts or attributes of a given stimulus have meaning in themselves). The relational cognitive style, on the other hand, requires a descriptive mode of abstraction and is self-centered in its orientation to reality; only the global characteristics of a stimulus have meaning to its users, and these only in reference to some total context. (Cohen 1969:829–30)

For obvious reasons the analytic style is clearly correlated with success in the academic context. While they are perhaps not identical, the analytic style is certainly close to what Jensen refers to as level II learning ability. Jensen attributes learning differences to genetic factors; Cohen rejects the genetic hypothesis and substitutes one in which socialization and group structure constitute the independent variables in the formation of cognitive style. In my opinion, her arguments are more convincing than Jensen's.

Observation indicated that relational and analytic cognitive styles were intimately associated with shared-function and formal styles of group organization. The manner in which critical functions were distributed in them seemed to parallel closely the observable cognitive functioning of their members. When individuals shifted from one kind of group structure to the other, their modes of group participation, their language styles, and their cognitive styles could be seen to shift appropriately to the extent that their expertise in using other approaches made flexibility possible. It appeared that certain kinds of cognitive styles may have developed by day-to-day participation in related kinds of social groups in which the appropriate language structure and methods of thinking about self, things, and ideas are necessary components of their related styles of group participation and that these approaches themselves may act to facilitate or impede their "carriers'" ability to become involved in alternate kinds of groups. (Cohen 1969: 831)

Methods of thinking about group process were very similar to methods of approaching test items. For example, it was functional for members of shared-function groups to perceive themselves to have relevance only within the contexts of their social groups, and they demonstrated on tests of conceptual style that they perceived parts of standardized stimuli to have relevance only when viewed within their given contexts. They did not extract and specialize functions in their primary groups, and they were not led to extract and organize parts of standardized stimuli on tests. The reverse was true of polar-analytic pupils. Style of primary-group participation appeared to have produced a subtle component of learned behavior that could be viewed in both the development of its related conceptual style and in light of its utility in subsequent group processes. (835)

Attention was then directed to nonverbal methods of measuring intelligence and achievement. Nonverbal tests have been designed to reduce their culture bias by drastically reducing their information components. However, they were found instead to deal in a focused fashion with the demonstration of analytic conceptual skills. Rather than freeing themselves of their culture-bound characteristics, they have focused on one critical aspect of it—the analytic mode of selecting and organizing information. When contextual inputs have been held constant, then, non-verbal tests of intelligence are more discriminatory against relational pupils than are the conventional types, which test partly for information growth. (842)

In a review of intelligence tests and heredity Fuller and Thompson (1960) compared a series of studies designed to separate genetic effects on various capabilities. Bluett found that tests of verbal fluency, verbal comprehension, and reasoning tended to be influenced by heredity to a greater degree than number and space abilities. Thurstone, Thurstone, and Strandskov (1953) found that spatial relationships showed strong hereditary influence, and reasoning the least. On other factors these scholars were in general agreement with Bluett.

Newman studied 50 pairs of dizygotic (two-egg) twins. Of these, three pairs were reared apart. In the DZ twins the average I.Q. differences were 9.9 points. The MZ (monozygotic) twins differed by 5.9 but the MZ's reared apart were on the average 9 points apart. In 1937 Newman, Freeman, and Holzinger studied 19 pairs of MZ twins reared apart as well as a series of

MZ and DZ twins reared together. The MZ apart sample varied in age from eleven to thirty-five years and one pair was fifty-three at the time of the study. Nine of the twins were separated at the end of one year, six during the second year, two during the third year, one at 6½ years, and one at the age of 8. Differences in the amount of education between members of each pair were also considerable. Intra-pair correlations reveal the following data.

	MZ Twins		
	Together	Apart	DZ
Binet	.910	.670	.640
Otis	.922	.721	.621
Stanford Achievement	.955	.507	.883

The results appear to show both genetic and hereditary effects. The only reversal occurs on the Stanford Achievement test. This can be explained on the basis of differences in education, an important environmental factor. When we compare studies of identical twins with those which attempt to demonstrate hereditary differences between "racial groups," two important facts must be emphasized. (1) The testing of individuals demonstrates both hereditary and social factors in the concept of intelligence as it is defined. I think that these results can be accepted as both biologically and sociologically valid because they are performed on units which are, in fact, clearly biological in nature, i.e., individual organisms. Group testing, on the other hand, has never been controlled for biological factors and is therefore equivocal. The only sure differences which can be revealed are social (this does not mean that biological differences are not also revealed, only that we cannot separate them from the social). (2) More attention should be paid to the responses which lie behind test performance. Such analysis in rat strains has contributed greatly to our understanding of the actual processes that go into producing the final response. Fortunately we do have a few studies of human performance which tap some of these dimensions, particularly as they involve sociological variables, in reference to the performance of sociologically defined "racial" groups.

Hess (1955) among many others has suggested that culture-free tests are necessary for the objective testing of intelligence. Such tests had the following requirements:

(1) The problems should represent as closely as possible real life situations rather than academic tasks;
(2) The problems, vocabulary employed in instructions, materials, and motivation should offer no cultural advantage to any group tested. (p. 21)

On the basis of such a test:

Individual differences were discriminated . . . and ability to predict reading achievement test scores within socioeconomic status groups was approximately equal to that of standard measures of intelligence. . . . The results emphasize the influence of cultural factors in test performance and suggest that socio-economic differences between high and low status samples in this country are exaggerated by standard intelligence tests. (p. 58)

Brigham, who had done considerable research on intelligence and had been convinced that differences between immigrant groups could be objectively tested, offered this strong renunciation of his own past theoretical bias:

This review has summarized some of the more recent test findings which show that comparative studies of various national and racial groups may not be made with existing tests, and which show, in particular, that one of the most pretentious of these comparative racial studies—the writer's own—was without foundation. (Brigham 1930:165)

Such candor is rare even in the annals of science. Brigham offers the following analysis for different responses to the alpha test. Note how his discussion parallels the analysis of behavioral responses in animal studies.

The assumption is made that people taking the alpha test adopted two different attitudes or sets, viz., a "directions attitude"—an attitude of careful attention to the examiner's instructions without looking at the test questions while the directions were read; and a "reading attitude"—partially or completely ignoring the examiner's instructions while studying the test questions during the time in which the examiner was reading. The adoption of the first attitude would tend to give the individual higher scores in test 1 (entirely oral directions), test 6 (an unusual form of mathematical test), and in 7 (a novel type of verbal test). On the other hand a person adopting the second attitude might quickly find out what was required in tests 3, 4, 5 and 8, and his score would be better if he ignored what the examiner was reading and studied the test questions during the period of instruction. (pp. 162–63)

This is the first indication that the problem of constructing a "culture-free" test is not the only one in intelligence testing. The procedures themselves may have an effect on the responses of the individuals.

Nonetheless, the argument continues. Jensen (1969), citing heritability studies of I.Q. as well as studies of white-Negro differences, has emphasized the genetic factor in intelligence. He suggests that .80 of the variance found among individuals on I.Q. tests can be attributed to genetic factors. Taking this heritability estimate as the basis of discussion, Jensen argues that of the average difference between groups of whites and blacks tested (usually 15 points of I.Q.) only about 7 points can be attributed to environmental factors, the rest being due to heredity. The Jensen report created quite a stir. The article is long, difficult, and replete with statistics. These factors have apparently impressed many people. Coming at a time when there is a concerted attempt to reopen questions concerning intelligence and race, particularly in the South where opponents of desegregation have taken new hope from the Nixon administration's so-called southern strategy, Jensen's paper attracted immediate attention.

While I shall attempt to show why I think Jensen's paper fails to prove a relationship between intelligence and races, I should first like to call attention to a point which appears to have been overlooked by most of those on both sides of this issue. Suppose, for a moment, that Jensen is completely right. What could this seven to seven and one-half points of intelligence mean in terms of educability? It should be clear that the suggested difference is so small as to mean very little, if anything.

Several cogent criticisms of Jensen's argument have

been offered. (See Bodmer and Cavalli-Sforza 1970: 19–29.) Crow, one of the leading population geneticists, finds much of value in Jensen's analysis, particularly his treatment of heritability, but when it comes to group differences, the crucial argument in the Jensen controversy, he offers an important caveat.

Heritability studies have been confined almost exclusively to white populations and to highly normal environments. How relevant are they to other populations and environments? We are currently especially concerned about culturally disadvantaged groups and racial minorities. Strictly, as Jensen mentions there is no carry-over from within-population studies to between-population conclusions.

I agree that it is foolish to deny the possibility of significant genetic differences between races. Since races are characterized by different gene frequencies, there is no reason to think that genes for behavioral traits are different in this regard. But this is not to say that the magnitude and direction of genetic racial differences are predictable.

It is clear, I think, that a high heritability of intelligence in the white population would not, even if there were similar evidence in the black population, tell us that the differences between the groups are genetic. No matter how high the heritability (unless it is one) there is no assurance that a sufficiently great environmental difference does not account for the differences in the two means, especially when one considers that the environmental factors may differ qualitatively in the two groups.

. . . It can be argued that being white or being black in our society changes one or more aspects of the environment so importantly as to account for the difference. For example, the argument that the American Indians score higher than Negroes on I.Q. tests—despite being lower on certain socioeconomic scales—can and will be dismissed on the same grounds: some environmental variable associated with being black is not included in the environmental ratio. (Crow 1969:307–8)

This criticism totally vitiates Jensen's arguments touching on black-white differences, for if they rest on a faulty set of comparisons we need take no other steps in a counter-argument. Nonetheless, there are other weak points in Jensen's analysis worth noting. For example, the statistical manipulation which is employed to demonstrate the heritability of I.Q. differences between whites and blacks can be turned around to indicate that all differences are due to environmental effects. This has been demonstrated in an unpublished paper by Rosedith Sitgreaves entitled "Comments on the Jensen Report." Sitgreaves, a statistician, offers a mathematically valid model which contradicts Jensen's results. Let me make it clear that this paper does not invalidate Jensen's argument but merely demonstrates that statistical theory alone cannot be used to buttress a theory of heritability.

Professor Jensen believes that the best estimate of H_1, derived from available data on correlations between I.Q. scores of individuals with varying degrees of family relationships, including monozygotic twins reared apart, is about .80. However, as he points out, this is a sample value, based on calculations made from relatively small samples, in which the sampled environments may not be represented in the same proportions as in the norming population, so that the true value of H_1 is still in doubt. For example, in the Bert study of 53 identical twins reared apart, 29 of the natural parents and 32 of the foster parents were in semi-skilled or unskilled occupations, so that the occupational distributions of the parents are skewed, and concentrated on one end of the occupational scale.

Suppose, however, in the time being, that we accept the value of .80 as the value of H_1. In such a case some simple calculations give us

$$G_1 = 100, \quad V_{G_1} = .80 \cdot (200) = 160,$$
$$E_1 = 0, \quad V_{E_1} = .20 \cdot (200) = 40$$

[*where 200 = the variance*] for the norming population. That is, a white population with a normal distribution of environments. The corresponding standard deviations are S.D.$_{G_1}$ = 12.6 S.D.$_{G_1}$ = 6.3.

Now let us consider a second population, namely a Southern Negro population. A number of studies have shown . . . that for this group the mean I.Q. score is about 85 while the standard deviation is about 12.6. That is, we have $P_2 = G_2 + E_2 = 85$, and $V_{P_2} = V_{G_2} + V_{E_2} = 160$. The values of the component means and variances depend upon the joint probability distribution of G and E in the Negro population.

The hypothesis proposed for study by Professor Jensen states that the observed differences in the means and variances of the I.Q. scores reflect differences in the distribution of the genetic component in the two groups. The alternative hypothesis proposed here considers that this distribution is the same in both the white and Negro populations with the result that $G = G_2 = 100$, and $V_{G_1} = V_{G_2} = 160$. Now, if we assume that for Negroes in the South, the totality of the environmental effects, represented by the component E, is to depress each I.Q. score by a fixed amount, and we assume that this amount is 15 points, we have $E_2 = -15$, $V_{E_2} = 0$, so that $P_2 = G_2 = E_2 = 100 - 15 = 85$, and $V_{P_2} = V_{G_2} + V_{E_2} = 160 + 0 = 160$.

Thus we obtain from the model exactly the values that have been observed. (Sitgreaves 1969:3–5)

Jensen's estimates of heritability are based upon studies of twins, particularly those studies concerned with twins who were reared apart. In addition to taking the highest heritability figure available, Jensen relies upon mean values rather than the observed range of I.Q. differences. While the mean values are relatively low, the range may, in fact, be quite wide. Since the available sample is quite small, a wide range of variation might be quite significant. In addition, as Sitgreaves points out, the environments of many of the twins who were reared apart are not significantly different. According to Gottesman:

Another way to gain perspective about the meaning of a 10 or 20 point IQ point difference is to look at the data on within-pair differences in intelligence for identical (MZ) and fraternal (DZ) twins. . . . Even though the gene pools do not differ and even though each of the two groups has been raised under more or less the same regime, the mean difference amounts to 6 IQ points for the sample of fifty pairs studied by Newman *et al* (1937). The range of within-pair differences was 0 to 20 points. Thus, even when gene pools are *known* to be matched, appreciable differences in mean IQ can be observed that could only have been associated with environmental differences.

A better appreciation of the influence of the environment on IQ can be gained from looking at the two unique samples of thoroughly described identical twins who have been reared apart and thus in discriminably different environments . . . the average intrapair difference on the Binet was 8 IQ points. The range of differences was 1 to 24 points. A very similar picture is given in a remarkably large sample of thirty-eight pairs of identical twins reared apart and studied by Shields (1962). When the tests used in this larger study are converted into IQ point equivalents (Shields and Gottesman, 1965) the average interpair difference for the identicals is 14 points on a verbal IQ test and 10 points on a nonverbal test. . . . At least 25 percent of the sample of identicals reared apart had within-pair IQ point differences exceeding 16 points on at least one of the tests . . .

It is obvious from looking at the data on identical twins that individuals with exactly the same genetic constitution can differ widely on the phenotypic trait we measure with IQ tests and label intelligence. The differences observed so far between whites and Negroes can hardly be accepted as sufficient evidence that with respect to intelligence, the Negro American is genetically less endowed. (Gottesman 1968:27–28)

It is apparent from his article that Jensen accepts race as a valid biological division (at least in the statistical sense) lying between the species and the population, at a time when there is growing evidence from statistical and taxonomic studies of within-species variation that race is an inappropriate concept in biology, particularly in regard to man. Let me remind the reader that this suggestion, first made in anthropology by Frank Livingstone and C. Loring Brace, has been borne out by Jean Hiernaux (1968) in his most recent book, *La diversité humaine en Afrique subsaharienne*. Hiernaux argues that, far from being uniform, African populations reveal a diversity comparable to that of populations from other regions of the world. Hiernaux demonstrates that African populations are no more homogeneous than any other group of geographically defined human populations.

In addition, Hiernaux has shown that groupings based upon ethnic classification in Africa do not correspond with those few genetic groupings above the level of the individual population which he was able to abstract from the data. In fact, of 101 ethnic groups open to cluster analysis, Hiernaux was able to establish only twenty-five constellations, of which twenty-three comprised only two populations. Forty-nine populations escaped all grouping.

While Jensen is talking specifically about the American Negro, the genes he is talking about (or, more correctly, a good percentage of them) come from Africa. Thus, some analysis of ethnic and genetic variation in Africa is germane to the discussion.

Gottesman (1968), in a book edited by Jensen and others, discusses the geographic range of populations in Africa from which slaves were imported to Charleston during the period 1733 to 1807. His figures, taken from a study by William Pollitzer, show the following percentages: Senegambia 20 percent; Winward Coast 23 percent; Gold Coast 13 percent; Whydah-Bennin-Calabar 4 percent; and Angola 23 percent. (Gottesman 1968:17) Such a distribution covers more than a thousand miles of coastline. In addition, the populations from which these slaves were drawn covered territory extending for six hundred miles inland. The range of genetic and ethnic groups tapped was extensive, to say the least.

In the United States itself, it is a tremendous over-simplification to speak of a single black or white population. According to Gottesman, "The variation observed in the studies reviewed in this section are probably valid and reflect the genetic heterogeneity of Negro Americans living different geographical and social distances away from their white neighbors. Such heterogeneity prevents us from speaking validly of an 'average Negro American' with x percentage of white genes." (Gottesman 1968:20)

Conclusions attributing a genetic basis to intelligence differences between groups are, to say the least, premature. To cite Gottesman again, "At the present time Negro and white differences in general intelligence in the United States appear to be primarily associated with differences in environmental advantages." (46)

In sum, genetic studies of black vs. white intelligence (whatever that is) which are based upon undifferentiated United States samples are naive in the extreme because they do not consider distributions of genetic variation in Africa or in the United States.

Jensen finds an overall intelligence deficit of 15 percentage points among American Negroes. He is willing to attribute about half of this difference to environment. The other 7½ points are assumed to reflect genetic factors. Yet on page 100 of his article he states:

In addition to these factors, something else operates to boost scores five to ten points from first to second test, provided the first test is really the first. When I worked in a psychological clinic, I had to give individual intelligence tests to a variety of children, a good many of whom came from an impoverished background. Usually I felt these children were really brighter than their IQ would indicate. They often appeared inhibited in their responsiveness in the testing situation on their first visit to my office, and when this was the case I usually had them come in on two to four different days for half-hour sessions with me in a "play therapy" room, in which we did nothing more than get better acquainted by playing ball, using finger paints, drawing on the blackboard, making things out of clay, and so forth. As soon as the child seemed to be completely at home in this setting, I would retest him on a parallel form of the Stanford-Binet, a boost in IQ of 8 to 10 points or so was the rule; it rarely failed, but neither was the gain very often much above this. So I am inclined to doubt that IQ gains up to this amount in young disadvantaged children have much of anything to do with changes in ability. They are largely a result simply of getting more optimal conditions . . . I would put very little confidence in a single test score, especially if the child is from a poor background and of a different race from the examiner.

Is Jensen not aware that these conditions are not met by the majority of studies he cites, particularly those drawn together by Shuey (1966)? If the deficit he notes is consistent in disadvantaged children, then all the IQ differences noted between blacks and whites in the United States may be subsumed under a combination of testing errors and environmental effects.

The Jensen paper contains other distortions and misinformation concerning cited data. The following material was extracted by Ms. Carole Vance and myself from a close reading of the Jensen Report and original sources.

On page 23 of the Jensen Report the author refers to an article by Burt (1963). He says that in the general Negro population there is an excess of IQs in the 70–90 range (see Jensen's illustration on page 25 of the report). This excess is explained as the combined effects of severe environmental disadvantage and emotional disturbance, which act to depress test scores. On page 27 Jensen says that Burt corrected for this bulge by eliminating scores of those having depressing factors. However, according to Burt's article there is a lack rather than an excess in the 70–90 range (cf. figure 1 in Burt 1963:180).

On pages 40–41 of his article Jensen cites Cooper and Zubek (1958). He stresses the effects of rearing bright rats in normal and enriched environments and says, "While the strains differ greatly when reared under 'normal' conditions . . . they do not differ in the least when reared in a 'restricted' environment and only slightly in a 'stimulating' environment."

In our opinion Cooper and Zubek put things the other way around and stressed the benefit of stimulation to dull animals. "A period of early enriched experience produces little or no improvement in the learning ability of the bright animals, whereas dull animals are so benefited by it that they become equal to bright animals. On the other hand, dull animals raised in a restricted environment suffer no deleterious effects, while bright animals are retarded to the level of the dulls in learning ability." (Cooper and Zubek 1958:162)

If one compares Jensen's figure 6 on page 50 of his report with figure 1 of Erlenmeyer-Kimling and Jarvik's (1963) article from which Jensen's data is drawn we find that Jensen shows only the midpoints for correlations between relatives reared together and reared apart. This emphasis stresses the discreteness and difference among the correlational scores while the original diagram, which shows the range and the median, demonstrates the overlap of correlational range and hence the overlap of strength of genetic inheritance.

On page 63 Jensen cites a study by Wheeler (1942) of IQ among Tennesee mountain children and notes that environmental improvements do not counteract a decline in IQ of "certain below average groups." Jensen neglects to mention Wheeler's discovery that the decline in IQ is due to the large percentage of left-back children. That is, a factor that raises the age level in every grade depresses the IQ scores. When Wheeler separates out the scores of older children in each grade he finds that the other children perform normally. Comparing chronologically "true" members of each grade over time (with the over-agers weeded out) he finds that in most years there is no decline. Wheeler says that the chronological IQ drop of 20 points is accounted for by children being repeatedly left back, which means more older children will be found as the grades get higher. Their presence depresses IQ scores most in the higher grades. If we follow Wheeler's rationale the decline which Jensen presents ranging from 103 to 80 points of IQ is reduced to 102.76 to 101.00.

On page 74 of his article Jensen says that on the average first-born children are superior mentally and *physically* to their siblings. His citation for this is Altus (1966). Altus, however, presents no evidence about physical superiority. Altus does cite a study by Huntington showing differences in birth order and achievement which suggests that differences are caused by superior physical strength of the first-born. Altus has the following to say about Huntington's hypothesis: "While his finding is typical of all those reported thus far, his explanation of the linkage is *not* typical: He argued that the first born probably tend to be physically stronger and healthier . . . *One may safely accept his data on the birth order of the eminent without accepting his explanation.*" (Altus 1966:45) (Italics ours)

On page 76 of his report Jensen cites Burt's (1961) contention that inheritance of intelligence conforms to a Mendelian, polygenetic model. Yet he does not note the wide variety of intelligence within a class and the fact that children's scores are not as narrow as those of their parents. In fact if there were no social mobility at all and classes were totally static the result of breeding over five generations would be a disappearance of class-means. "After about five generations the differences between the class-means would virtually vanish, and the proportional range within each class would spread out almost as widely as the proportional range of the population as a whole. (Burt 1961:15)

British studies show that class IQ scores have been remarkably stable over the past hundred years. This is because social mobility has moved up bright lower-class children and moved down less bright upper-class children. Burt's study appears to bear this out for England.

Now, if we apply this model to the United States Negro, intelligence would have remained constant by class if social mobility operated as in England. But even in the lowest class, there would be children of above average intelligence who would rise so that the range of child intelligence would be much wider than adult intelligence (regression to the mean). In any case Jensen does not mention this aspect of Negro performance, that is, unexpected over-performance.

This model cannot be applied to the United States, however, because there is little real social mobility for Negroes. Since most would be held in the lower classes one would expect an evening out of IQ scores within five generations, that is, if one is to follow Burt's model.

On page 83 of his report Jensen cites research by Heber and Dever on education and habilitation of the mentally retarded. While we do not have Jensen's original source for this research (a paper read at the Conference on Sociocultural Aspects of Mental Retardation), we do have a paper by Heber and Dever entitled "Research on Education and Habilitation of the Mentally Retarded," appearing in *Social-Cultural*

Aspects of Mental Retardation, edited by H. C. Haywood, New York, Appleton-Century-Crofts, 1969.

Jensen says that Heber has estimated that IQs below 75 have a much higher incidence among Negro children than among white children at every level of socioeconomic status. (Jensen, p. 83)

We find no statement that Negroes have a higher frequency of IQs under 75 than whites. Furthermore, Heber's study was not a study of race and intelligence but rather a study of a special group of mentally retarded from a specific neighborhood in Milwaukee described as follows:

Characterized by having the city's highest known prevalence of mental retardation among school age children. The nine census tracts which comprise this area, known as the "inner core", also has the city's highest rate of dilapidated housing, the greatest population density per living unit, the lowest median income level, and the greatest rate of unemployment. Though comprising no more than five percent of Milwaukee's population it yields about one-third of the mentally retarded known to the schools. (Heber 1969:35)

The point of Heber's study is to show how mental retardation is probably cultural and not genetic.

On page 86 of his paper Jensen cites a study by Heber (1958) which discusses precocity of Negro infants. Jensen mentions motor precocity but neglects to mention intellectual development as well. Heber says, "The results of the tests showed an all round advance of development over European standards which was greater the younger the child . . . The precocity was not only in motor development; it was found in intellectual development also." (Heber 1958:186)

Furthermore, in a long section running from page 194 to page 195 of his article, Heber presents an environmental explanation for the observed precocity.

For those of us who have done field work in Africa, there can be little doubt that behavioral and cognitive patterns in the black community in the United States bear little resemblance to those in Africa and that, in fact, such patterns are highly variable in Africa and dependent more upon cultural differences than upon shared gene pools. Hiernaux's data on the lack of correspondence between ethnic and genetic identity are of direct relevance here.

The main thrust of Jensen's paper, which has been somewhat buried by popular accounts, is actually that there is a wide diversity of mental abilities in men and that there is a need to develop diversified educational programs. With this I concur heartily, and it is unfortunate that Jensen's paper has influenced some to attack this very point. By seizing upon the genetic hypothesis without an awareness that genetic potential is manifested only in relation to environmental experience, we lose all the gains achieved by our growing understanding of the interaction of heredity and environment. I should like to think that Jensen is himself committed to further experimentation in education. It is unfortunate that he published what can only be taken as a premature analysis of the genetic effect on intergroup differences. Jensen has taken a fairly safe hypothesis—that intelligence is heritable—and forced it to carry the burden of a second argument for which there is still little evidence: that black and white performance on intelligence tests is determined primarily by genes.

Two recent studies (Cohen's work cited above is also relevant) have amplified the role of culture and

social group in both test results and academic performance. Katz (1968) varied test conditions for samples
of Negro students in relation to "subjective probability
of success"—that is, how the individual taking the test
felt about the likelihood of his own success. Differences
in this attitude were then measured against different
types of testing situations in which the race of the tester
was varied as well as the kind of attitude expressed
during the testing situation. The theoretical basis behind these studies is that of need achievement in which
"the strength of the impulse to strive for success on
a given task is regarded as a joint function of the person's motives to achieve, the subjective problems of
success and the incentive value of success." According
to the model, on a test that has evaluative significance
(e.g., a classroom test) motivation is maximal when
the probability of success is at .50 level.

Katz notes that in a number of experiments with
Negro college students, such individuals tend to under-
perform on intellectual tasks in the presence of whites.
Katz speculates

that for Negroes who find themselves in predominantly
white academic achievement situations, the incentive value
of success is high but the expectancy of success is low
because white standards of achievement are perceived as
higher than own-race standards. By the same token, the
perceived value of favorable evaluation by a white adult
authority is high, but the expectancy of receiving it is low.
Therefore, by experimentally controlling Negro subjects'
expectancy of success on cognitive tasks it should be possible to produce the same, if not higher, levels of performance in white situations as in all-Negro situations. (Katz
1968:134)

A group of Freshmen were given a test which was described to them as part of a scholastic aptitude test. They were told that their scores would be evaluated in comparison to scores achieved in predominantly white colleges. The students were given a pretest and then told what their chances of success on the actual test would be. One-third were led to believe that they had little chance of meeting the standards for their group, one-third were told they had an even chance and one-third were told they had a good chance. Each of these three groups was then divided into sub-units, one given a white tester, the other a Negro tester.

The results showed that in the low and intermediate probability conditions, performance . . . was better with a Negro tester, but when the stated probability of achieving the white norm was high, the performance gap between the two tester groups closed. (Katz 1968:134)

Another test, in which a college with no admission standards other than high school graduation was compared to a college with high relative standards, demonstrated that the effects of varying the race of the tester was the same as in the controlled experiment described above. On the other hand, the scores achieved by students at the selective college were higher when the testers were white, no matter what the probability of success.

The differences were explained by Katz as follows:

In summary, it appears that Negro students who had been average achievers in high school (the nonselective college sample) were discouraged at the prospect of being evaluated by a white person, except when they were made to believe that their chances of success were very good. But

Negro students with a history of high academic achievement (the selective college sample) seemed to be stimulated by the challenge of white evaluation, regardless of the objective probability of success. (Katz 1968:135–36)

Katz generalizes the results of these studies in terms of differences in the socialization process between lower- and middle-class children. "The present assumption is that lower class children, . . . because they have received less parental approval for early intellectual efforts *remain more dependent than middle class children on social reinforcement when performing academic tasks.*" (Katz 1968:138)

To relate this generalization to interaction with white and Negro adults a group of Northern urban Negro children of elementary school age were tested on a list of ten paired associates for ten trials. Anxiety was reduced by assuring the subjects that the results would not affect their school grades. The children were divided into two initial groups, one tested by whites the other by Negroes. Within each group one-half received periodic reinforcement in terms of praise and one-half were discouraged through verbal disapproval of their performance. The sample was also dichotomized in relation to need for approval, which was measured for all subjects before the experiment was run.

More learning occurred with Negro testers than with white testers and more learning occurred when the testers supported the students' efforts.

When the Negro tester was approving, all boys, regardless of need level, were adequately motivated for the task. When the Negro tester was disapproving, high-need boys were somewhat disheartened but continued to seek approval. . . . Low-need boys, on the other hand, tended to lose interest in the task, since it was defined at the outset as having no academic significance. When the tester was white and gave approval, high-need boys did not work quite as hard as boys who received approval from the Negro tester. . . . Though the white person's approval was seen as less genuine, their high need generated a moderate impulse to work (perhaps, in part, to avoid disapproval). Low-need boys were relatively unmotivated by white approval. When the white tester was disapproving, high-need children experienced debilitating anxiety, because the disapproval was taken as an overt expression of dislike; it was as though they could not hope to elicit a favorable response through greater expenditure of effort. These boys were virtually unable to learn at all. When low-need subjects were disapproved of by the white adult their performance did not deteriorate further. (Katz 1968:141–42)

While these experiments do not relate directly to intelligence testing they go a long way toward explaining why certain sociological groups respond as they do to the educational process. The problem is complicated since it involves the motivation of the individual, which is in part a product of his home experience as well as his conception of the expectations of teachers defined partially in terms of their race. The common educational experience of lower-class Negro children with white teachers in school situations which are often discouraging, and in the context of environmental settings in which probability of success is lowered by the experience of daily life, probably has an effect on all types of test performance, particularly intelligence tests.

The process of learning in children is even more

subtle than Katz's impressive findings would indicate. A recent study of performance of school children in the San Francisco area supports the hypothesis that the teacher's attitude toward the probable success of the child will have a profound effect on the outcome of the educational process. This effect may be profound even when the teacher is unaware of specific elements of his own performance in the learning situation.

The study in question begins with a series of interesting findings relating to learning in our (by now familiar) pure strain rats. Students in psychology were given rats to run in a simple learning experiment. Half of them were told that the rats were bright; the other half were told that the animals were from a dull strain. The rats which were labeled as bright lived up to their expectations and outperformed their dull brothers. Remember that these rats were all of more or less equal ability. The test outcome could only have been the result of bias on the part of the experimenters. It is important also to point out that it was the performance of the rats which differed, not the students' evaluation of that performance.

Armed with these results, the experimenters then turned to the school community. The experimenters established the expectation in teachers that certain children in the school chosen at random would show superior performance in the coming school year. This expectation was established by testing the children on an intelligence test and misinforming the teachers of the results. The use of this test in the pre-experimental situation had the added advantage of providing a measure since the children could be re-examined with the same test during the course of the experiment. A casual method of informing the teachers about the existence in their classes of "Potential spurters" was used. "The subject was brought up at the end of the first staff meeting with the remark 'By the way, in case you're interested in who did what in those tests we're doing for Harvard.'" (Rosenthal and Jacobson 1968:22)

All the children were retested four months after school started, at the end of the school year, and finally in May of the following year. As the children matured, they were given tests appropriate to their level. These tests were designed to evaluate both verbal skills and reasoning.

The results indicated strongly that children from whom teachers expected greater intellectual gains showed such gains. . . . The tests given at the end of the first year showed the largest gains among children in the first and second grades. In the second year the greatest gains were among the children who had been in the fifth grade when the "spurters" were designated and who by the time of the final test were completing sixth grade.

At the end of the academic year 1964–1965 the teachers were asked to describe the classroom behavior of their pupils. The children from whom intellectual growth was expected were described as having a better chance of being successful in later life and as being happier, more curious and more interesting than the other children. There was also a tendency for the designated children to be seen as more appealing, better adjusted and more affectionate, and as less in need of social approval. (Rosenthal and Jacobson 1968:22)

The results are interesting in themselves but an unanticipated finding of the study has even more startling impact. An interesting contrast became apparent when teachers were asked to rate the undesignated children. Many of these children had also gained in I.Q. during the year. The more they gained the less favorably they were seen!

The most unfavorable ratings were given to the children in low-ability classrooms who gained the most intellectually. . . . Even when the slow-track children were in the experimental group, where greater intellectual gains were expected of them, they were not rated as favorably with respect to their control-group peers as were the children of the higher track and medium track. Evidently it is likely to be difficult for a slow-track child, even if his I.Q. is rising, to be seen by his teacher as well adjusted and as a potentially successful student. (Rosenthal and Jacobson 1968:22)

There is some evidence that bright children who are understimulated in the classroom tend to become discipline problems. In middle-class neighborhoods such behavior is sometimes perceived as a sign of intelligence. This is not the case in the experimental school described by Rosenthal and Jacobson. If a teacher has low expectations of an "active" student, the student is defined as a troublemaker. When this is the case, the self-fulfilling prophesy contributes to the creation of socially and educationally maladjusted children.

These studies contribute to our understanding of the dynamics of learning and the testing situation. They call into question the results of previous tests which have attempted either to prove or disprove the general superiority of one biological or sociological group over another. Such investigations highlight once again the long-known difficulty of constructing a culture-free test, one which is neither dependent upon specific learning nor upon differences in motivation among those groups being compared. In addition, these studies reveal differences between sociologically and psychologically defined groups in individual motivation, self-perception, and the potential effect of testers and teachers upon the performance of those being tested. It is clear that the manipulation of psychological and sociological variables can produce almost any type of result. The evidence therefore calls into question the validity of all biological interpretations of those experiments which have established group differences. The differences are probably real enough, but the strong likelihood is that they were due in all cases to sociological rather than to biological factors. This conclusion is inescapable when we add to it the fact that the groups compared in all cases owe their reality to sociological definitions of race rather than to some sound biological criteria. The whole process of comparing "racial" groups on the basis of intelligence has been misguided by the confusion which exists between "race" as a valid sociological concept and "race" as a biological concept.

A recent study of intelligence *testers* tends to show that the final interpretation of data is highly correlated with certain biographical characteristics of the investigators themselves! Thus, studies of comparative group intelligence may serve as better predictors of the inherent prejudice of psychologists and sociologists than of any real differences *or* similarities between groups.

Sherwood and Nataupsky (1968) found 128 authors who had contributed scholarly articles on intelligence research. Addresses were found for 104 of them.

Questionnaires concerning biographical material were sent to all 104 with the explanation that the authors were conducting an investigation of the personal characteristics of a selected sample of American scientists. Subjects were asked to respond anonymously to 35 items, among which were the following:

1. Age when research was published
2. Birth order in family
3. Mother American or foreign-born
4. Mother's education
5. Father American or foreign-born
6. Father's education
7. Father's occupation and prestige
8. Grandparents American or foreign-born
9. Childhood of investigator, urban or rural
10. Integration of religion into family as child
11. Size of high school graduating class
12. Scholastic standing in high school
13. Nature of undergraduate college
14. Enrollment of college
15. Undergraduate major
16. Percent of undergraduate expenses earned
17. Age at Bachelor's degree
18. Years between B.A. and Ph.D.
19. Religious preference
20. Frequency of attendance at religious services
21. Membership in voluntary organizations

Eighty-two respondents returned the questionnaire, representing 64 percent of the original sample of authors. Of those contacted, 78 percent responded. Later biographical information on a sample of those who could not be located or who did not respond was collected from printed biographical sources with no significant differences in experimental outcome.

Seven dependent variables based upon intelligence test interpretations were delimited. These were:

1. Differences exist but are due to environment
2. Study not conclusive but indicates that the environment is the major variable causing differences
3. Negroes superior on some tests, whites on others
4. No differences between Negroes and whites
5. Study not conclusive because samples were not equated
6. Study not conclusive but there is an indication of innate inferiority of Negroes
7. Negroes innately inferior

These were then correlated with the biographical data. Seven biographical variables discriminated among these categories. These were: age at which research was published, birth order, mother's education, father's education, grandparents American or foreign-born, childhood rural or urban, and scholastic standing at college.

An overall test of significance was highly positive, with a *P*. of .001 well in the accepted range of statistical analysis.

Those who believed that intelligence differences were innate tended to be first-born. Those whose grandparents were foreign-born tended to see no differences between whites and Negroes while those whose grandparents were American by birth indicated definite findings of innate inferiority in Negroes. Those whose parents had low mean years of education saw environment as a crucial factor in test responses. Individuals from rural backgrounds were more likely than those whose origin was in urban centers to find indications of biological differences. Biological differences were also stressed by those who had higher scholastic standing as undergraduates. Those researchers who fell into groups 6 and 7 (indication of Negro inferiority or definite indication that Negroes are inferior) tended to be youngest in publication of research, first-born, with American-born grandparents, with parents who had many aggregate years of schooling, of rural backgrounds, and with high scholastic standing as undergraduates.

From this data the authors infer that those researchers who found indications of Negro inferiority came from higher socioeconomic backgrounds.

It would appear from these findings that not only is it impossible to make judgments about group differences in intelligence on the basis of published studies but that such studies serve as better predictors of testers' characteristics than of those tested! In retrospect such tests then do have a double sociological function. When carefully analyzed, they reveal sets of psychological and sociological variables which differentially affect the performance of sociologically defined "racial" groups. And their interpretations tend to predict certain biographical characteristics of those who design and analyze them. Thus, they tell us a great deal about the working of American society even if they tell us little about inherent differences between biological groups!

BIBLIOGRAPHY

Alland, A., Jr. (1967). *Evolution and human behavior.* New York, Natural History Press (Doubleday & Co., Inc.)

Altus, W. D. (1966). Birth order and its sequelae. *Science,* 151:44–59.

Baker, P. (1960). Climate, culture and evolution. In *The process of ongoing human evolution,* edited by Gabriel W. Lasker. Detroit, Wayne State University Press. Pp. 3–16.

Birch, H. (1968). Boldness and judgment in behavior genetics. In *Science and the concept of race,* edited by M. Mead, T. Dozhansky, E. Tobach and R. E. Light. New York, Columbia University Press. Pp. 49–58.

Blum, H. (1961). Does the melanin pigment of human skin have adaptive value? *Quarterly Review of Biology,* 36:50–63.

Bodmer, W. F., and L. L. Cavalli-Sforza (1970). Intelligence and race. *Scientific American,* 223:19–29.

Bogardus, E. S. (1925). Measuring social distance. *Journal of Applied Sociology,* 9:299–308.

—— (1933). A social distance scale. *Sociology and Social Research,* 17:265–71.

Boyd, W. C. (1963). Four achievements of the genetical method in physical anthropology. *American Anthropologist,* 65:243–52.

Brace, C. L. (1964). A nonracial approach towards the understanding of human diversity. In *The concept of race,* edited by Ashley Montagu. Glencoe, Ill., The Free Press. Pp. 103–52.

——, and A. Montagu (1965). *Man's evolution.* New York, Macmillan.

Brigham, C. C. (1930). Intelligence tests of immigrant groups. *Psychological Review,* 37:158–65.

Bruell, J. H. (1965). Inheritance of behavioral and physiological characteristics of mice and the problem of heterosis. In *Readings in animal behavior,* edited by T. E. McGill. New York, Holt, Rinehart and Winston. Pp. 126–142.

Buettner-Janusch, J. (1959). Natural selection in man: the ABO (H) blood group system. *American Anthropologist,* 61:437–56.

Burt, C. (1961). Intelligence and social mobility. *Brit. J. Stat. Psychol.,* 14:3–24.

—— (1963). Is intelligence distributed normally? *Brit. J. Stat. Psychol.,* 16:175–90.

Cavalli-Sforza, L. L. (1969). "Genetic drift" in an Italian population. *Scientific American,* 221:30–37.

Cohen, R. (1969). Conceptual styles, culture conflict, and nonverbal tests of intelligence. *American Anthropologist,* 71:828–56.

Coon, C. (1962). *The origin of races.* New York, Alfred Knopf.

—— (1965). *The living races of man.* New York, Alfred Knopf.

——, S. M. Garn and J. Birdsell (1950). *Races, a study of the problems of race formation in man.* Springfield, Ill., Charles C. Thomas.

Cooper, R., and J. Zubek (1958). Effects of enriched and restricted early environments on the learning ability of bright and dull rats. *Canad. J. Psychol.,* 12:159–64.

Cravioto, J., E. R. Delicardie, and H. G. Birch (1966). Nutrition, growth and neurointegrative development: an experimental and ecological study. *Pediatrics,* 38, No. 2, Part II, 319–72.

Crow, J. (1970). Genetic theories and influences: comments on the value of diversity. *Harvard Educational Review,* 39:301–9.

Erlenmeyer-Kimling, L., and L. F. Jarvik (1963). Genetics and intelligence: a review. *Science,* 142:1477–79.

Fuller, J. L., and W. R. Thompson (1960). *Behavior genetics.* New York, John Wiley.

Gajdusek, C. (1964). Factors governing the genetics of primitive human populations. *Cold Spring Harbor Symposium on Quantitative Biology,* XXIX:121–36.

Garn, S. (1965). *Human races.* 2d ed. Springfield, Ill., Charles C. Thomas.

Ginsburg, B., and W. Laughlin (1966). The multiple basis of human adaptability and achievement: a species point of view. *Eugenics Quarterly,* 13:240–57.

——, and W. Laughlin (1968). The distribution of genetic differences in behavioral potential in the human species. In *Science and the concept of race,* edited by M. Mead, T. Dobzhansky, E. Tobach and R. E. Light. New York, Columbia University Press.

Goodman, M. (1963). Man's place in phylogeny of the primates as reflected in serum proteins. In *Classification and human evolution,* edited by Sherwood Washburn. Chicago, Aldine Press. Pp. 204–34.

Gossett, T. (1963). *Race: the history of an idea in America.* Dallas, Southern Methodist University Press.

Gottesman, I. I. (1968). Biogenetics of race and class. In *Social class, race, and psychological development,* edited by Martin Deutsch, Irwin Katz, and Arthur R. Jensen. New York, Holt, Rinehart and Winston. Pp. 11–51.

Gregory, R. L. (1966). *Eye and brain.* New York, McGraw-Hill.

—— (1968). Visual illusions. *Scientific American,* 219:66–76.

Harris, M. (1964). *Patterns of race in the Americas.* New York, Walker and Co.

Heber, Rick F., and Richard B. Dever (1969). Research

on education and habilitation of the mentally retarded. In *Social-Cultural Aspects of Mental Retardation*, edited by H. C. Haywood. New York, Appleton-Century-Crofts.

Hess, R. D. (1955). Controlling culture influence in mental testing: and experimental test. *Journal of Educational Research*, 49:53–58.

Hiernaux, J. (1964). The concept of race and the taxonomy of mankind. In *The concept of race*, edited by Ashley Montagu. Glencoe, Ill., The Free Press. Pp. 29–45.

—— (1968). *La diversité humaine en Afrique subsaharienne*. Editions de l'Institut de sociologie Université Libre de Bruxelles.

Hirsch, J. (1968). Behavior-genetic analysis and the study of man. In *Science and the concept of race*, edited by M. Mead, T. Dobzhansky, E. Tobach and R. E. Light. New York, Columbia University Press. Pp. 37–48.

Howell, C. (1952). Pleistocene glacial ecology and the evolution of "classic Neanderthal" man. *Southwestern Journal of Anthropology*, 8:377–410.

Huxley, J. (1942). *Evolution: the modern synthesis*. New York, Harper and Brothers.

Jelinek, J. (1969). Neanderthal man and *Homo sapiens* in Central and Eastern Europe. *Current Anthropology*, 10:475–503.

Jensen, A. (1969). How much can we boost I.Q. and scholastic achievement? *Harvard Educational Review*, 39: 1–123.

Katz, I. (1968). Some motivational determinants of racial differences in intellectual achievement. In *Science and the concept of race*, edited by M. Mead, T. Dobzhansky, E. Tobach and R. E. Light. Pp. 132–48.

Kelso, J. (1962). Dietary differences: a possible selective mechanism in ABO blood group frequencies. *Southwestern Lore*, 28:48–56.

——, and G. Armelagos (1963). Nutritional factors as selective agencies in the determination of ABO blood group frequencies. *Southwestern Lore,* 29:44–48.

Kilham, P., and P. H. Klopfer (1968). The construct of race and the innate differential. In *Science and the concept of race,* edited by M. Mead, T. Dobzhansky, E. Tobach and R. E. Light. New York, Columbia University Press. Pp. 16–25.

LeGros Clark, W. E. (1964). *The fossil evidence for human evolution.* University of Chicago Press.

Lindzey, G. (1967). Behavior and morphological variation. In *Genetic diversity and human behavior,* edited by J. N. Spuhler. Chicago, Aldine Press. Pp. 227–40.

Livingstone, F. (1958). Anthropological implication of sickle cell gene distribution in West Africa. *American Anthropologist,* 60:533–62.

—— (1960). Natural selection, disease, and ongoing human evolution, as illustrated by the ABO blood groups. In *The process of ongoing human evolution,* edited by G. Lasker. Detroit, Wayne State University Press. Pp. 17–27.

—— (1964). On the nonexistence of human races. In *The concept of race,* edited by Ashley Montagu. Glencoe, Ill., The Free Press. Pp. 46–60.

Marshall, G. (1968). Racial classifications: popular and scientific. In *Science and the concept of race,* edited by M. Mead, T. Dobzhansky, E. Tobach and R. E. Light. New York, Columbia University Press. Pp. 149–164.

Montagu, A. (1960). *An introduction to physical anthropology.* Springfield, Ill., Charles C. Thomas.

—— (1967). *Man's most dangerous myth: The fallacy of race.* Cleveland, Meridian Books.

Mourant, A. E. (1954). *The distribution of human blood groups.* Oxford, Blackwell.

Murdock, G. P. (1959). *Africa.* New York, McGraw-Hill.

Otten, C. (1967). On pestilence, diet, natural selection and the distribution of microbial and human blood group antigens and antibodies. *Current Anthropology,* 8:209–26.

Putnam, C. (1961). *Race and reason: a Yankee view.* Washington, D.C., Public Affairs Press.

Rees, Linford (1960). Constitutional factors and abnormal behavior. In *Handbook of abnormal psychology,* edited by H. J. Eysenck. New York, Basic Books. Pp. 344–92.

Rosenthal, R., and L. F. Jacobson (1968). Teacher expectation for the disadvantaged. *Scientific American,* 218:19–23.

Sahlins, M. (1968). La première société d'abondance. *Les Temps Moderne,* 24:641–80.

Scott, J. P., and J. L. Fuller (1965). *Genetics and the social behavior of the dog.* University of Chicago Press.

Shapiro, H. (1960). *The Jewish People,* UNESCO.

Sheldon, W. H., S. S. Stevens, and W. B. Tucker (1940). *The varieties of human physique.* New York, Harper and Brothers.

Sherwood, J. J., and M. Nataupsky (1968). Predicting the conclusions of Negro-white intelligence research from biographical characteristics of the investigator. *Journal of Personality and Social Psychology,* 8 (part 1): 53–8.

Shuey, A. (1966). *The testing of Negro intelligence.* New York, Social Science Press.

Sitgreaves, R. (n.d.). Comments on the Jensen report. Paper read at the Meetings of the National Academy of Education, 1969.

Spuhler, J. (1959). *The evolution of man's capacity for culture.* Detroit, Wayne State University Press.

——, and G. Lindzey (1967). Racial differences in behavior. In *Behavior genetic analysis,* edited by Jerry Hirsch. New York, McGraw-Hill. Pp. 366–414.

Stanton, W. (1960). *The Leopard's spots: Scientific attitudes toward race in America, 1815–1859.* University of Chicago Press.

Thurstone, T. G., L. L. Thurstone, and H. H. Strandskov (1953). *A psychological study of twins.* Chapel Hill, The

Psychometric Laboratory, University of North Carolina.

Turnbull, C. (1961). *The forest people,* New York, Simon and Schuster.

Wagatsuma, H. (1967). The social perception of skin color in Japan. *Daedalus,* 96:407–43.

Washburn, S. (1963). The study of race. *American Anthropologist,* 65:521–31.

Wheeler, L. R. (1942). A comparative study of the intelligence of East Tennessee mountain children. *J. educ. Psychol.,* 33:321–34.

INDEX

C